SO-CFV-821

ideas in Science

CONTEMPORARY PHYSICS

DAVID PARK

Williams College

A HARBINGER BOOK

Harcourt, Brace & World, Inc.
New York   Chicago   Burlingame

COVER PHOTOGRAPH: Red light issuing from a gas laser is deflected by a prism. The light rays are very nearly parallel and would diverge only a few inches in a mile of travel. (Courtesy Perkin-Elmer Corporation, Norwalk, Conn.)

F.6.70

© 1964 by Harcourt, Brace & World, Inc.

All rights reserved. No part of this publication may be reproduced or transmitted in any form or by any means, electronic or mechanical, including photocopy, recording, or any information storage and retrieval system, without permission in writing from the publisher.

Library of Congress Catalog Card Number: 64–23458

Printed in the United States of America

ISBN 0-15-622566-2

Preface

This book is written as personal experiment, to see whether I can interest a general reader who does not wish to be reminded of technicalities in a subject which abounds in them. Those directly involved in physics spend most of their time on details which have a life and interest of their own, but the wonderful spread and flow of the general scientific ideas emerging from all this work is what gives it permanent value.

My attempt is to set out conclusions while being sketchy about the ways in which they were reached, and even about the proper terms in which they should be described. This may seem pointless and even unscientific to colleagues to whom the chief beauty of science lies in the methods used. But if I can with my generalities and simplifications reach some intelligent readers who would not ordinarily linger over a scientific book, and persuade them of the intellectual importance and excitement of the work that physicists are doing, now, over much of the world, I shall consider that to be justification enough.

These chapters originated as lectures given at Williams College in the spring of 1961 at the kind invitation of the Society of Alumni. They were taken down on tape. Several revisions have, I hope, expunged most traces of the oral presentation. I have about doubled the amount of material and tried to bring it up to date. My thanks must go to the Williams alumni, who sat through the original lectures, to Mr. Edwin Ripin, who suggested the book and helped greatly in preparing the text, and to Mr. John Bremer for his critical comments.

DAVID PARK

Williamstown, Mass.
June 1964

Contents

Contemporary Physics

CHAPTER *1*

Mathematical Philosophy

The main purpose of this book is to show a general reader some of the developments in the science of physics during the last few years. To those involved in it, the story has been a fascinating one of darkening confusion in some areas and enlightenment, sometimes very sudden, in others. There is certainly much more known about nature than there was a generation ago, and the new knowledge extends both outward and downward—outward to include newly discovered facts and downward to include new insights into the fundamental nature of physical reality. But before I can hope to communicate much of this, I must first try to explain what physics is and how it differs from other sciences.

The difficult point is that many sciences overlap, both in their subject matter and in their methods of approach. However, to stress this fact too much would give a misleading impression of chaos. There are, to be sure, a *Journal of Chemical Physics* and a *Journal of Physical Chemistry,* but the general viewpoints of the articles in the two are noticeably different, so that it is fairly clear in most cases in which of the two a given article belongs.

Physics is the most fundamental of the sciences, and in a way the simplest. The biologist studies the structures and functions of living organisms and relates them

1

to the properties of the individual cell, an enormously complicated organization of matter that he would like to understand. The biochemist helps by making a study of the complex chemical compounds that a cell contains. The basic structural unit of such compounds consists of an immense number of atoms, whose arrangement the science of chemistry is gradually making clear. The physicist is interested in individual atoms, and his interest may stop there or extend to atoms linked together in twos and threes. Beyond these limits he usually studies a gigantic number of atoms, perhaps 10^{23} or so,* arranged regularly as in a crystal or chaotically as in a gas. In such cases the object of his study will usually be to show how some aspects of crystals in general, or of gases in general, can be understood in terms of the properties of the individual atoms composing them.

Deeper than this are questions like "Why are atoms as they are?" or "Why are there some kinds of atoms and not others?" These are answered in terms of the structures of atoms—the smaller particles of which they are composed and the laws governing their behavior. Then the same two questions are asked about the smaller particles.

Finally there are deeper questions still to which all this work leads naturally: "What is the nature of human knowledge of the external world?" and "Are there any definite limits to what we can know?" Most physicists are interested in questions like these and have opinions on them, but since it is difficult to ask them precisely, or even to know when one is talking sense about them, only a few hardy souls have seriously devoted themselves to such matters.

To define a science is to limit it, and this we must not attempt to do. Physicists, for example, have recently made profound contributions to the understanding of living cells, and they did much of the early work on the theory of aircraft flight. But the disciplines begun in this way soon take on lives of their own. They develop separate names (in these cases, microbiology and aerodynamics), corps of specialists, and

---

*The meaning of the notation 10^{23} is discussed in the Appendix.

their own scientific journals, while physics continues in its search for fundamentals. The notion of physics that will be developed here is that of a systematic explanation of natural phenomena, which proceeds from a knowledge of structure on the one hand and laws of motion on the other, on the principle that if in some way one can find out how a certain section of the universe is composed and how its component parts behave, he can then infer and explain its function.

NATURE AS A MECHANISM

The description of physics just given would apply very well to the act of explaining how a piano or an automobile engine works. The physicist, however, is not interested in these devices for their own sakes. His scientific curiosity extends beyond them to objects much larger, much smaller, and much more mysterious. Often, as in his studies of atomic phenomena, he finds that he is not able to observe directly either the structure or the function of what he is studying and that he must subsist entirely on indirect hints derived from measurements whose very interpretation often poses the most severe problems. To work out a detailed theory from such hints, he uses a method that is not used by the piano tuner or the automobile mechanic—the method of mathematics. Mathematics plays a crucial role in the entire structure of physical science as we know it, so much so that it would be quite impossible to imagine contemporary physics stripped of its mathematical content. Mathematics is used in physics because it is a system of relationships by means of which one can draw exact conclusions. This does not mean that you can understand a thing from a few hints just by using enough mathematics. You can't make something from nothing no matter how clever you are. The function of mathematics is rather to make general ideas as specific as they can be made, so that a physicist who has an idea can express this idea unambiguously, see what its consequences are, and, hopefully, see whether the consequences check with something that is definitely known or that can be found out.

Mathematics is a device for unraveling complex relationships that goes far beyond what intuition is capable of, and it has been found to be almost miraculously effective in constructing the science of physics. This "unreasonable effectiveness of mathematics," as it has been called, is probably due to the fact that much of the world's mathematics has been invented by physicists looking for an appropriate tool; therefore, we select for our study those physical systems that are capable of being understood in mathematical terms and restrict ourselves to the consideration of questions that can be answered by mathematical methods. There is no guarantee, however, that we shall continue to be so lucky in the future. It is possible to imagine a nonmathematical physics arising someday (as it did in Aristotle's work, for example), but as yet there is no sign of it, and since the eighteenth century treatises on physics have occasionally referred to themselves as books on mathematical philosophy.

The essential role that mathematics plays in physical thought is most clearly felt by a physicist just after he has been to a lecture on public affairs. The facts are fairly familiar to everybody, the reasoning of the lecturer is clearly set out, and yet the hearer may depart totally unconvinced by what he has heard. It does no good that the lecture may have been supported by figures and statistical reasoning; these are isolated bits of mathematics set into a framework of argumentation that is not mathematical in nature and that is not necessarily convincing. The great merit of a correct mathematical argument in physics is this: that it *is* necessarily convincing to anyone who has mastered the techniques of mathematics. If the reader of a mathematical exposition in physics will grant the premises, and if he follows through the mathematics without finding any errors in it, then he must grant the conclusion. (The actual situation is not quite this neat, since it is generally impossible to be sure what all the premises are!) It is unfortunate that no one has found either a way of rephrasing discussions outside the scientific field so that they can be conducted by contemporary mathematical methods or of extending mathematics to the point at which it could deal with the kind of questions that ordinary people ask themselves in the modern world. The Royal Society in England concerned itself with this problem in the eighteenth century with rather funny results, which can be recognized from Swift's satire on them in the Laputa episode in *Gulliver's Travels*.

The word *mathematics* covers an immense variety of techniques but for our purposes mainly the differential calculus of Newton and the integral calculus of Leibniz. (These attributions are a little too pat, for the development of calculus from algebra took a century, and both of these men were familiar with both kinds of calculus.) Calculus deals with relations. Algebra deals with numbers. A relation such as $2a - 3b = 0$ assigns to any number a another number b and vice versa (i.e., if $a = 3$, then $b = 2$). One puts in a number and gets a number out. Calculus deals with the relation itself. One puts in one relation and gets out another one. Relations are what interest physicists most, and this is why calculus is the language in which most physics is naturally expressed.

The most basic type of mathematical relation in physics is that known as a law. "The distance from the earth to the moon is about 240,000 miles" is not a law, though it is true, because it is not a general statement. There are many moons in the solar system, and the statement touches only one of them. "Every time two rough surfaces are rubbed together, heat is developed" is not a law either, because although it is general, it gives no idea how much heat results from how much rubbing. A passable law can be expressed by giving the relation between the two quantities, but it is only a special case of a more general law, that whenever a certain amount of one form of energy disappears, an exactly equal amount of energy in other forms takes its place. This is the law of the conservation of energy, and it covers the example just mentioned, in which mechanical energy becomes heat energy, as only a small part of its total scope.

Some of the physical laws we know have been inferred from measuring what happens in many individual events. Some are derived mathematically from other laws. Others, like the conservation of energy, were at first the hard-won fruit of experience but can now be derived mathematically. Is there a set of fundamental laws of nature from which all others can be derived? We do not know. Probably the question is meaningless anyhow, for lack of a definite standard of "fundamentalness." But the intellectual goal of physics is to find more and more specific facts and to explain them by fewer and fewer basic principles.

A law is a relation between numbers. The output of a physics laboratory is mostly in terms of numbers. The view of such a laboratory in the comics and in motion pictures is often highly qualitative, with flashing lights, beautiful colors, sparks leaping back and forth, and mysterious sounds. These effects provide entertainment after a busy day, but they can only entertain, because they are phantoms that cannot be grasped by mathematics. Physics and mathematics make contact through numbers, and, in the final analysis, the observations recorded in the laboratory must be translated into numbers before the process of reasoning that distinguishes the physical approach can be applied. Thus the physicist works his way through a tangle of relationships, but his final result is a number.

How, exactly, are general physical laws made to yield specific knowledge? Let us consider the phenomena of astronomy as seen through a telescope: a vast number of tiny points of light, all of them in motion, and some, the planets, moving differently from the rest. We must first order our sensations by asking certain questions. What are we looking at through the telescope? Is the night sky, as some of the ancients believed, a black sphere with holes in it through which light streams from an illuminated region beyond? Are the points of light self-luminous bits of matter moving through the air? Or (as has turned out to be the case) are they very large masses of hot substance very far away in empty space? Such different ways of picturing the natural basis of our sensations are known as *models,* and the physicist starts his work by constructing in his mind a model to explain the phenomena he has in front of him.

The reader will recall that the late Renaissance was

marked by a struggle between proponents of two alternative models of the astronomical universe, that of Ptolemy and that of Copernicus. In the Ptolemaic model (Figure 1–1) the earth stands stationary at the center of the universe, and a complex mechanism whirls the stars, the planets, and the sun around it. In the Copernican model (Figure 1–2) the sun is at the center, the stars are stationary, and the planets, including the earth, move around the sun. If one starts with one of these two models, he can draw certain conclusions from it, which he can then test by observation. It is easy to see that if the models are taken perfectly literally, without regard to their implications, and if the stars are very far away, there is no way to distinguish by observation the universe of Ptolemy from that of Copernicus. To an observer standing on the earth, they appear exactly the same. But it turned out that the Copernican model is vastly more fertile than the Ptolemaic model in suggesting mathematical explanations of astronomy. The Copernican model was the basis of Isaac Newton's interpretation of how the planetary system operates. According to Newton, the sun is a center of gravitational attraction, and the planets move around it in roughly circular orbits because they are attracted to it and kept by this gravitational force from escaping from its influence and moving off into space. The existence of the gravitational attraction can be verified by experiments performed in the laboratory. Although the Ptolemaic model also had a mechanical explanation (in terms of turning crystalline spheres), this was purely *ad hoc* and bore no direct relation to any other scientific experience.

The simplest model of the solar system, and the one with which Newton began his work, assumes that the sun is extremely massive and that the planets are very much less massive by comparison—so much so that the only gravitational forces that come into play are those between the sun and the different planets, those between one planet and another being negligible. Now, the law of universal gravitation says that *every* two bodies in the universe attract each other by a gravitational force. Thus, what we are saying here is mani-

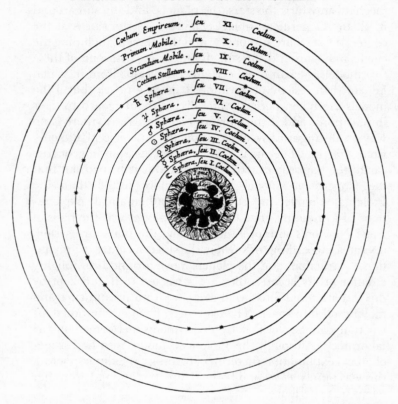

FIGURE 1–1   Model of the Ptolemaic system. (From
O. von Guericke, *Experimenta Nova Mag-
deburgica de Vacuo Spatio*, Johann Jansson,
Amsterdam, 1672.)

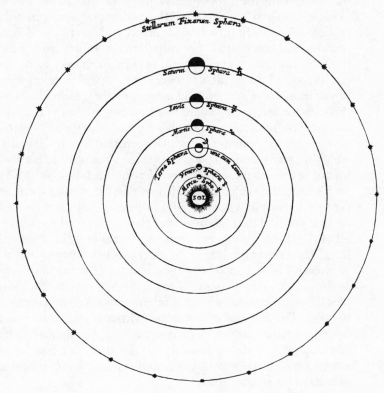

FIGURE 1–2 Model of the Copernican system. The fifth orbit is that of the earth with the moon. No other moons had been discovered by 1543. (From *Ibid.*)

festly untrue and, in fact, contradicts the law of gravity according to which it was set up. The advantage of the model, however, outweighs its disadvantage, for it is easy to study mathematically the way in which a system of astronomical objects having the properties of those in the model would behave. Newton investigated the system with the mathematical methods that he had invented and the physical laws that he had discovered. He calculated numbers giving the positions that the planets should occupy in the sky at different times, and the result was a very nearly perfect description of the motions of planets and comets as they were observed through the telescope. There were, to be sure, some discrepancies. The planetary motions were not exactly as Newton had calculated, but we have just seen that the model with which he was working did not correspond exactly to his own theory of the solar system. Once Newton had shown the way by his simplified calculations, it was possible during the next century to take into consideration the small interactions that he had omitted and to show that they accounted for nearly all of the discrepancies between his figures and the observed facts. The point here is that Newton was not being primitive or wrong. He was merely being scientific. It is not necessary at the outset to give attention to all the fine details of a complex physical system, and it is often confusing and pointless to do so. The first task of a scientist with a new idea is to find out whether he is correct in a general way; he can do this by envisioning a model of an object under study, simplifying it where it needs to be simplified, and trying to understand it as a basis for further study.

What distinguishes a good model from a bad one? A good model has two outstanding characteristics. First, it explains a wide variety of observed facts. Second, it is simple enough, and presents itself to the imagination sufficiently clearly, that it is an abundant source of new ideas. The Ptolemaic system of astronomy was, of course, simple enough in its basic outlines, but it gave undue importance to the earth in the scheme of astronomical things and led to a series of unscientific notions of man's place in the universe that have had to be extensively revised. No real understanding of the physical world could possibly have come out of the Ptolemaic system, whereas the Copernican system has already transformed our ideas of the nature of man and of the world he inhabits and is still richly productive.

Our concept of nature as a mechanism expresses itself through the creation of models with general features that can be intuitively grasped and with detailed properties that can be worked out by mathematical means. That this technique succeeded so well for the solar system has greatly influenced the intellectual climate of the last two centuries. It has contributed to a philosophical outlook called rationalism, which implies that, subject to our inherent limitations of intelligence, we can, if we work long enough and hard enough, arrive at a mechanically causal and (very likely) mathematical explanation of everything we observe. Much has happened since the eighteenth century to shake this view and to show that, if it is to survive at all, it must be modified. But we can never go back to the feeling of wonder and helplessness in the face of natural forces that belonged to the ages preceding it.

CLASSICAL PHYSICS

The physics of Newton and his successors, which is now commonly called classical, is largely the physics of mechanical models of things. The starting points are therefore first a model and second the general physical principles assumed to underlie its function. Newton invented many of the ideas of classical physics. His explanation of the solar system, which was the science's first great success, depended first on the law of universal gravitation, which he discovered, and second on the laws of motion, to which he was led somehow by laboratory measurements.* He was able to show that these laws

---

* Newton's correspondence has been published (H. W. Turnbull and J. F. Scott, eds., *The Correspondence of Isaac Newton*, Cambridge University Press, 3 vols., 1959–1961), and work is slowly progressing on an edition of his surviving papers. These ought to tell us much more about his creative process than we now know.

could be applied without change to movements of the planets and satellites. The law of gravitation, to which we shall return in Chapter 4, states that every two objects in the universe attract each other by a force that is proportional to the mass of each object and inversely proportional to the square of the distance between them. That is, the force of gravity between the earth and the moon would be just twice as great as it is if the moon were twice as massive as it is, or if the earth were twice as massive as it is, and four times as great as it is if the earth and moon were separated by only half the distance they are. The laws of motion need not be described in detail at this point, but they enable one to calculate the subsequent movements of objects acted upon by any set of forces and started off in any particular way.

The third basic ingredient of classical physics is the idea of atoms. This idea, originating in the middle fifth century B.C.,* is at the very heart of Newton's science. The following passages, taken (in reverse order) from the end of his *Opticks,* illustrate the quality of his thought; the questions raised in the second paragraph are still far from settled.

All these things being consider'd, it seems probable to me, that God in the Beginning form'd Matter in solid, massy, hard, impenetrable, moveable Particles, of such Sizes and Figures, and with such other Properties, and in such Proportion to Space, as most conduced to the End for which he form'd them; and that these primitive Particles being Solids, are incomparably harder than any porous Bodies compounded of them; even so very hard, as never to wear or break in pieces; no ordinary Power being able to divide what God himself made one in the first Creation. While the Particles continue entire, they may compose Bodies of one and the same Nature and Texture in all Ages: But should they wear away, or break in pieces, the Nature of Things depending on them, would be changed. Water and Earth, composed of old worn Particles and Frag-

---

* The atomic theory is about contemporaneous with the Parthenon and the plays of Sophocles and Euripides. The interested reader can form some idea of this time by looking at the sculpture and paintings it produced.

ments of Particles, would not be of the same Nature and Texture now, with Water and Earth composed of entire Particles in the Beginning. And therefore, that Nature may be lasting, the Changes of corporeal Things are to be placed only in the various Separations and new Associations and Motions of these permanent Particles; compound Bodies being apt to break, not in the midst of solid Particles, but where those Particles are laid together, and only touch in a few Points.

Have not the small Particles of Bodies certain Powers, Virtues, or Forces, by which they act at a distance, not only upon the Rays of Light for reflecting, refracting, and inflecting them, but also upon one another for producing a great Part of the Phaenomena of Nature? For it's well known, that Bodies act one upon another by the Attractions of Gravity, Magnetism, and Electricity; and these Instances show the Tenor and Course of Nature, and make it not improbable but that there may be more attractive Powers than these. For Nature is very consonant and comfortable to herself. How these Attractions may be perform'd, I do not here consider. What I call Attraction may be perform'd by impulse, or by some other means unknown to me. I use that Word here to signify only in general any Force by which Bodies tend towards one another, whatsoever be the Cause. For we must learn from the Phaenomena of Nature what Bodies attract one another, and what are the Laws and Properties of the Attraction, before we enquire the Cause by which the Attraction is perform'd. The Attractions of Gravity, Magnetism, and Electricity, reach to very sensible distances, and so have been observed by vulgar Eyes, and there may be others which reach to so small distances as hitherto to escape Observation; and perhaps electrical Attraction may reach to such small distances, even without being excited by Friction.

The atomic picture is fundamental to much of classical physics. It is the model used to explain a huge number of phenomena which, like the evaporation of a liquid, take place in a manner the senses do not follow. But it did not develop in any way during the time of Newton because, as we shall see later, to make a physics of atoms requires much more knowledge of how atoms are constructed and how they function than Newton could possibly have had. Still the general types of forces that we now know control the motions of atoms, as well as the motions of the electrons and nuclei of which atoms are composed, were known in the classical period. These are the forces of electricity and magnetism, mainly electricity, as Newton prophetically suggested at the end of the second passage just quoted.

The results of classical physics as they appeared at the end of the nineteenth century were very rich in some directions and remarkably poor in others. Planetary astronomy, as Newton visualized the subject—the explanation of the motions and apparent positions of objects in the heavens—was very nearly complete. The questions as to what and where a star really is, what it is made of, how it is constructed, and why it shines were, of course, barely considered in those days, and it is only recently that we have begun to have clear opinions about them. By the end of the nineteenth century everything having to do with purely mechanical principles had been analyzed thoroughly. The motions of fluids, for example, were very well understood, and great attention had been given to the types of movement that come under the general heading of waves. So much had been done with the theory of electricity and magnetism that the physicist of 1900 could predict mathematically the outcome of almost any reasonable electric or magnetic experiment involving wires, batteries, magnets, and measuring instruments. This work had enormous practical implications, and I need hardly stress that the by-products of the scientific thought of the classical period are largely responsible for the kind of civilization that we have today.

Generalizations from the analyses of particular models with which the classical period began showed by 1900 that the economics of the universe is organized around certain principles of barter or exchange, whose existence was not anticipated in the early days. The most important unit of this exchange is energy. The word "energy" in its modern sense was entirely unknown to physicists at the time of Newton. It is defined in textbooks as the ability to do work, which indicates at least something of its nature, for a physical object that contains energy in some form or other is always able to do work—that is, to move something that resists being moved. But this is not a very enlightening way of explaining the term, and the idea cannot really be understood except by experience with the many forms that energy can take, for it can transform itself like the Old Man of the Sea, and the progress of modern physics can to some extent be mapped by the successive discoveries of different ways in which the ability to do work is hidden or disguised in nature. The most obvious and notable examples of energy are the energy of motion, contained by a moving object that does work of various kinds as it is being brought to rest, and the energy of radiation, which brings us light and warmth from the sun. Einstein in his early work on relativity demonstrated that mass and energy are the same, that a thing has energy if it has mass and mass if it has energy. The two quantities are connected by a universal relationship, the famous $E = mc^2$, where E is the energy, m the mass, and c the speed of light.

One of the most striking results of nineteenth-century physics was that the total amount of energy in the world probably remains fixed, even though it changes from one form to another. The acceptance of this principle was gradual and rests entirely on the ability to measure energy in its different forms. Such a principle is called a law of conservation, and because it can be stated in such very general and simple terms, it occupies a central position in physical thought. Other laws of conservation are known, notably the law of conservation of electric charge, which states that although electricity can be moved around, and by the use of modern methods made to manifest itself in different forms of matter, there is no way to alter, to the slightest degree, the total amount of electric charge in the world.

Newton and his intellectual successors visualized the whole world in terms of atoms. Yet it was not until the end of the nineteenth century that the nature of atoms began to be understood. One can guess from Newton's writings that, having explained astronomy with a model in which forces act upon objects and these objects obey laws of motion, he wished to turn his attention to chemistry and explain it in exactly the same way; probably the greater part of Newton's scientific activity was directed toward this end. His years of chemical experimentation came to nothing, however, and he was entirely unsuccessful in identifying the forces acting between atoms and establishing their behavior.

Newton's errors are apparent from his idea of a gas. The most characteristic feature of a gas is its tendency to expand, which suggested to Newton that the atoms of a gas repel each other. He visualized a gas as consisting of atoms rather widely strung out in space, perhaps vibrating around their equilibrium positions, and pushing outward from each other so that if a hole were made in a containing vessel, the atoms nearest the hole would be forced out through it. Only gradually was this model superseded by another, the kinetic model. The kinetic model has predominated for the last century, and it asserts that the fundamental consideration in understanding the nature of a gas is that the atoms are by no means stationary or even approximately stationary in space but that they move freely except for occasional collisions among themselves and with the containing walls. In this view, the pressure that a gas exerts is the result of the impacts of its atoms against the walls and against one another, which occur millions of times every second. (In most gases, the moving particles are not atoms but molecules, each consisting of two or more atoms tightly bound together by chemical forces, but this is only a detail.) Of the two gas models, the second has survived and forms the basis of what is now called the kinetic theory of gases.

THE KINETIC THEORY

Feeling or looking at a sample of matter, one does not sense anything of its atomic nature; therefore, all the properties of atoms must be inferred from indirect evidence. The first and best quantitative evidence available on the two models of a gas was provided by experiments on heat, and in particular on the conservation of energy in transforming mechanical energy to heat and vice versa. Rubbing the hands together on a chilly day not only makes them feel warmer; they *are* warmer, because the mechanical energy dissipated in the rubbing reappears in the form of heat. The inverse transformation is more obvious—it can be seen in any steam or gasoline engine. According to the kinetic theory, there is no essential difference between heat energy and mechanical energy. Heat energy is the energy of motion of the molecules of which a hot substance is composed, and so it is mechanical energy. The only distinction is that in heat energy the motions are random and uncontrollable, because the molecules are so very small, whereas in a moving baseball or steam-engine shaft all the molecules proceed in the same direction. It is possible for a baseball as a unit to be at rest relative to us. It is also imaginable for all its molecules to be at rest relative to us, or at least to each other. In this condition, the baseball would have no heat energy in it at all. This hypothetical case suggests one of the central ideas of the subject of heat, that of an absolute zero of temperature. In the primitive nineteenth-century picture of the atomic theory, all molecular motion ceases at absolute zero, and this is the base from which all calculations of heat are made. The notion of an absolute zero is like that of an infinite straight line in geometry, and we shall see that it must be rendered somewhat more sophisticated before it corresponds very closely to reality. But it has been possible to locate absolute zero in terms of the ordinary scales of centigrade and Fahrenheit temperature (it is about $-273°C$ or $-459°F$) and to cool down very small samples of matter and measure their properties within one or two thousandths of a degree of it.

Arguments dealing with heat are not, however, conclusive proof of the atomic hypothesis as long as atoms cannot be seen, for heat is heat, and the kinetic theory of matter is essentially a mechanical one. In order to be convincing, physics had finally to produce an observable mechanical consequence of the action of atoms, and this was the first and almost the greatest of Albert Einstein's contributions. Before Einstein it was entirely possible for a very good scientist to believe that the atomic theory had no ultimate value at all as an explanation of nature.

Einstein sought a link between the mechanical motions of atoms and mechanical motions on a scale visible to human beings. At the beginning of the twentieth century, atoms were many thousands of times too small to be detected by any known physical means. Einstein's problem was to determine whether there existed any physical object large enough to be viewed through a microscope and still small enough to respond in some way to the mechanical actions of atoms. He found them in the phenomenon known to biologists as Brownian motion. If one suspends a fine powder such as plant pollen in a liquid and studies the grains through a microscope, he notes that they are in continual trembling, random motion, which does not die away with time. Once he has shown that this motion is not a manifestation of life or any other transient cause (it persists even after many years), he naturally asks whether it is not the result of atomic bombardment. In 1905 Einstein not only replied yes to this question but also gave a quantitative answer: the higher the temperature at which Brownian motion is observed, the more frequent and more vigorous the atomic bombardments are, and the more briskly a particle of pollen moves around. Einstein analyzed the motion mathematically and showed that although, of course, it takes place in a completely random manner, it is still governed by definite laws of probability. He was thus able to predict how the motion would depend on the grain size and the temperature. Here was an experiment explicable entirely in mechanical terms, starting with the model furnished by the kinetic theory of heat. Measurements in subsequent decades showed that Einstein was right.

At this point in human history reasonable people could no longer deny that the kinetic theory of matter had a foundation of truth. And although much was to happen before the finer details of the atomic hypothesis could be worked out, there was no doubt that the physics of the twentieth century would have to concern itself primarily with the structure and function of atoms.

CHAPTER *2*

The Atomic Picture of Nature

Newton proposed to explain the complexity of chemical phenomena by assuming that atoms were hard spheres with certain forces acting between them. The sizes of the spheres were presumed to differ, and perhaps also the kinds of forces. He wished to be able to account for the immense variety of chemical phenomena observable in the laboratory as a consequence of these differences. The kinetic theory of gases makes use of essentially the same assumption: the hard spheres are infinitely elastic and lose no energy when they bounce together; since an atom in the gas is moving around fairly far from its neighbors most of the time, the exact nature of the attracting forces between atoms is not very important. (A typical atom travels on the average about 200 times its own diameter between collisions, at some 500 miles per hour.) As it turned out, the kinetic theory of gases was a brilliant success, whereas the Newtonian theory of chemistry had practically no application. It seems that a concept involving small particles in flight, where it doesn't matter much what the exact physical nature of the particles is, can be somewhat loose and still be useful, whereas one involving chemical combinations in which the particles stick together in definite configurations that necessarily depend on their inner nature must be

19

much more specific. In fact, no progress could be made at all in atomic physics without some knowledge of the structure of atoms, and the nineteenth century had few techniques for acquiring such knowledge.

Any scientific study starts with experimentation. Thus the first problem confronting early investigators of atoms was to decide what was evidence for the nature of atomic structure and what was not, and to invent experiments that would interrogate the inner constitution of these extremely small particles. Chemistry was not of much use for this purpose because, although chemical reactions are very specific and can be duplicated in the laboratory, they are enormously varied, and it is not at all easy to see by looking at a chemical reaction what the underlying atomic process is. Therefore, simpler and more direct means were necessary. It was only in the last years of the nineteenth century that evidence leading to the modern view of atoms began to accumulate.

THE ATOM OF 1912

The way in which one investigates almost any object that can be studied in a laboratory is to subject it to stimuli and see how it responds. Atoms are no exception. The question is, what kind of stimuli are available and what kind of responses can one identify? Not many happenings on the ordinary laboratory level of size reflect directly what goes on inside an atom. But there is one thing that can easily be observed—the light that atoms give out when they are subjected to an excitation of some sort, let us say heat.

I would like to make an analogy between the emission of light by an atom and the emission of sound by a musical instrument. One musical instrument can be distinguished from another by its tonal quality. In more scientific terms, a musical instrument producing a certain note gives out not only that note but also a number of others, called overtones, and the ear is sensitive enough to identify and distinguish them. If all musical instruments gave out but one note, they would sound exactly alike. Similarly, when atoms are struck or otherwise excited, they give out light, and this light is of various wavelengths, which can be measured and distinguished. Unfortunately, the eye is virtually useless for this purpose because it contains no mechanism that can separate out and analyze different wavelengths of light as the ear can analyze different wavelengths of sound; however, an instrument called a spectroscope can do this for us.

Returning to our analogy, let us say that we are going to study the sound of a bell. In order to study the overtones produced when the bell is struck, we must hang it in a way that does not interfere with its free vibration. We then strike it and listen. The wrong way to study a bell would be to put it into a bag together with a lot of other bells and then shake the bag. The result would be a random and disorganized noise from which no conclusions could be drawn. Clearly the bell must be isolated from everything else and allowed to vibrate freely. The same requirement is necessary in studying the radiation from atoms. A white-hot piece of metal emits light at all frequencies, with no particular frequencies distinguishable or predominating. Such light emission is analogous to the noise that would issue from a bag of bells. Evidently the mechanical oscillations of an atom, whatever may be their nature, must be examined when the atom is not interacting with its neighbors; therefore, the atom is studied in the form of a gas. Although at atmospheric pressure the collisions between atoms in the gas are very frequent, if the pressure is reduced enough, they can be made relatively rare, and then the atoms begin to give out sharply limited radiations of light at certain definite wavelengths but not at any others. In fact, a trained spectroscopist can put a sample of a substance into his instrument and determine, merely by looking through the spectroscope at the light emitted by the substance's hot vapor, what chemical elements are present. This is the technique that enables astronomers to be so wise about the chemical constitution of stars, since much of the light that reaches us comes from their gaseous outer layers. (Our knowledge of stellar interiors is purely from mathematics.)

When different gases are investigated in this way, it is found that there is one whose spectrum has a distinctive property. This is hydrogen gas, treated rather specially so that the spectrum is due to the vibration of the individual hydrogen atoms in the gas and not to that of pairs of hydrogen atoms locked together in molecules. The spectrum of hydrogen as seen in Figure 2–1 is a regular pattern of wavelengths that

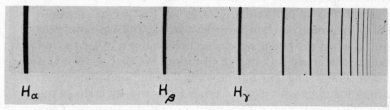

FIGURE 2–1   Spectrograph of the Balmer series of hydrogen. (From G. Herzberg, *Spectra of Diatomic Molecules*, 2nd ed., Van Nostrand, Princeton, N.J., 1950.)

repeats itself two or three times, and the different wavelengths observed can even be summed up in a single formula. No other gas has such a simple and orderly spectrum. The formula was found in 1887 by the Swiss schoolmaster Jacob Balmer and was for many years one of the curiosities of physics. It had no logic behind it but merely summarized all the spectroscopic data available on hydrogen atoms in a convenient equation that gave, with great accuracy, the wavelengths of all the known lines. For contrast, Figure 2–2 shows the more disorderly spectrum of helium, another light gas.

A completely different way of stimulating an atom so that one can observe its reactions is to bombard it with the fast-moving particles that emerge from a radioactive nucleus. Experiments along this line (Figure 2–3) were made in 1909 at the laboratory of the University of Manchester under the direction of Ernest Rutherford, and they led him to the conclusion that most, if not all, of the mass of an atom is concentrated in the center of it. The model that Rutherford proposed to demonstrate his conclusion was very much like the Copernican model of the solar system. The analogy

FIGURE 2-2   Spectrograph of part of the helium spec-
              trum. (From H. G. Kuhn, *Atomic Spec-
              tra,* Longmans, Green, London, 1962.)

is striking, for most of the mass of the solar system does reside
in its central member, the sun. Rutherford suggested such a
model in a paper in the *Philosophical Magazine.* The suggestion
was immediately taken up by Niels Bohr, a young research
student at the same laboratory, who set himself the task of
explaining the spectra of the atoms and, in particular, the
spectrum of hydrogen, which, judging from the existence of
Balmer's formula, ought to be easier to explain than any
other.

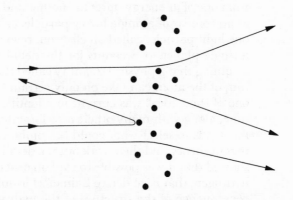

FIGURE 2-3   Rutherford's interpretation of the be-
              havior of fast particles directed at a thin
              metal foil. (A real foil is, of course, many
              thousands of atoms thick.) The occa-
              sional deflection through a large angle
              is explained as the result of an almost
              head-on collision with a nucleus that is
              small and massive.

It is not necessary to go into the details of Bohr's theory here, for they have been largely superseded. His main innovation was to show that atomic systems seem to be different from the objects normally encountered in a laboratory in that the quantities of internal energy which they may contain are limited to certain definite amounts. We are used to thinking that we can put any arbitrary amount of water into a glass until it is filled and that we can put any arbitrary amount of heat into a thermometer until it explodes. But we cannot put any arbitrary amount of energy into an atom; until a certain point is reached, the energy it can contain is restricted to certain levels. These energy levels can be represented diagrammatically as in Figure 2–4.

Bohr's theory assumes that each radiating atom gives out only one frequency at a time, the complexity of spectra being due to the simultaneous emission of light from many different atoms. It assumes that light is emitted only as an atom passes from one of its energy states to another and that the hydrogen atom consists of a single heavy particle, called a proton, with one light particle, called an electron, revolving around it like a single planet. It accounts for the definite energy states by assuming that a physical quantity called the angular momentum of the atom can take on only certain fixed values. Every one of these ideas was current in scientific discussions at the time, along with many others now forgotten; Bohr's achievement was to select what could be combined into a single coherent theory and then work out the details of the theory. He showed that it was possible to explain exactly the spectrum of hydrogen, that is, to derive Balmer's formula as the necessary consequence of the structure of the hydrogen atom and the laws he assumed to operate.

Evidently the simplicity of the hydrogen spectrum is related to the fact that there is only one electron present; more complicated atoms contain more and more electrons, which interact with each other as well as with the central particle and complicate the spectrum in the same way that the interactions between planets complicate the behavior of the solar system. Bohr explained not only the spectrum of hydrogen

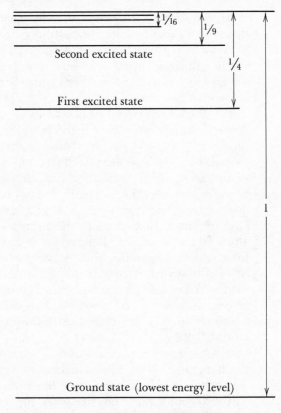

Second excited state

First excited state

Ground state (lowest energy level)

FIGURE 2–4   Energy levels of a single hydrogen atom.
Normally the atom is in its ground state.
If it is energized to a higher state, it emits
energy as a flash of light and regains the
ground state in roughly 10^{-8} second.

but also the size of its atom, known to be about 10^{-8} cm
(1/100,000,000th centimeter).* No formula was forthcoming
for the spectrum of helium or any of the other atoms, but

---

* Readers unfamiliar with exponential notation are referred to the Appendix.

Bohr was able to show that the chemical properties of the different elements follow beautifully from the planetary model which Rutherford had proposed and that the reasons for the immense variety of chemical properties met in nature are all to be found in the differing numbers of electrons that atoms can contain.

The Bohr-Rutherford picture of an atom heavier than hydrogen consists of a heavy particle at the center with very much lighter particles circulating outside it. The central particle has a positive electric charge and is called the nucleus of the atom. The circulating particles, or electrons, have negative charges. One of the basic laws of electricity is that positive and negative charges attract each other, and the force of this attraction keeps the electrons from leaving the atom. In a general way Bohr was able to show that not only the chemical properties but also some of the physical properties of many substances could be explained in terms of this model. A crystal, for example, turns out to be an array of heavy atoms, arranged in some orderly manner like bricks in a wall or squares on a chess board. Owing to the close proximity of other nuclei, some electrons may become detached from their own atoms and wander throughout the crystal. If this happens, it is possible to make the electrons flow through the crystal. In this manner a metallic substance conducts electricity. In other materials like glass, the wandering does not occur, so that the substance is an insulator. Bohr's theory could also explain a number of magnetic properties because of relations between electricity and magnetism that were known from studies made in the early nineteenth century. Thus within ten years of Rutherford's discovery, there was a general insight into the atomic nature of reality, which clarified many properties of atoms and their combinations and explained an impressive number of phenomena.

In spite of this success, there were signs that all was not well with the Bohr-Rutherford model. The most definite of these was that little progress had been made mathematically. Bohr and Rutherford assumed that the next most simple atom after hydrogen was helium, containing a doubly charged nucleus and two electrons circling around it. But at the end of a decade of the most strenuous mathematical investigation it could not be understood why a helium atom should be mechanically stable and not lose one of its electrons after a while, and even less why it gave out the particular pattern of wavelengths observed in its spectrum. Similarly, the simplest molecule—that consisting of two hydrogen atoms linked together—could not be understood in any mathematical way; again, this elementary example of a chemical force binding two atoms together was totally inexplicable, and it was impossible to see why the two atoms did not drift apart. To go any further in the understanding of atoms, it was necessary to develop completely new ideas. They will be discussed in the next chapter, where we will see that the Bohr theory of the hydrogen atom represents the extreme limit to which a mechanical model of a natural object can be taken. Here the classical version of physics runs out, and to advance from hydrogen to helium we must enter the modern era.

Nevertheless, the Bohr-Rutherford theory yielded a tremendous richness of new concepts that have completely transformed the physicist's view of nature, and this transformation is mirrored in the new terms that it introduced. He uses the word nucleus all the time, and the idea of elementary particles such as electrons and protons, with properties that can be studied by laboratory experiments, is basic to all of his work. Of the two elementary particles encountered in Bohr's theory, an electron is the smallest possible unit of negative electricity. Its mass is some two-thousandths that of a proton, which has an equal and opposite electric charge; a proton has a mass equal to about 2×10^{-24} gram. Both particles are now known to have a further remarkable property known as *spin*. It is easy to give a rough mechanical picture of what is meant by spin, but what actually happens does not have any exact mechanical analogy. It is as though the particles were little tops, but little tops of a strange variety that are always spinning and have only one possible rate of rotation. In this way their behavior suggests the situation described by Bohr, in which only certain values of energy are possible. Elementary particles, characterized by their mass, their charge, and their spin, will be discussed further in Chapter 6.

PROPERTIES OF NUCLEI

At the same time that Bohr's theory was exploring those properties of atoms that arise from their electronic structure, it was posing a whole new class of questions relating to the nucleus lying at the center of the atom, which it made no attempt to answer. The nucleus of a hydrogen atom is a single proton. What then is the nucleus of a helium atom? It must have a doubly positive charge in order to balance the charge of the two electrons circling it. One would therefore suppose it to consist of two protons in some way bound together. But here two quite separate difficulties arise. In the first place, it is possible to measure the mass of the helium nucleus, which turns out to be as great as that of four protons. In the second place, it is a law of physics that like electric charges repel each other, and two protons situated as closely together in space as they would have to be in a helium nucleus would repel each other so strongly that it is very difficult to see how they could stick together at all. The first puzzle was solved by the discovery in 1932 of a new elementary particle known as a neutron. It appears now that atomic nuclei consist of protons and neutrons together. The neutrons, as the name implies, have no electric charge and therefore make no contribution to the charge of the nucleus, but their mass is about that of a proton. The nucleus of the helium atom thus consists of two neutrons and two protons. All nuclei met with in nature can be characterized similarly. The discovery of neutrons solved a problem that had existed since 1911—the reason why there are different varieties of the same element that resemble each other in every chemical property except that the masses of the atoms are somewhat different. Atoms related in this way are called isotopes. Their explanation in terms of the Bohr-Rutherford model is simple; one need only assume that there is an extra neutron or so in the nucleus of the heavier isotope, which gives it its extra mass but which does not in any way affect the configurations of the electrons outside it. The second puzzle, the reason why a nucleus with more than one proton does not at once fly apart, took somewhat longer to figure out; it will be discussed in Chapter 7.

The Bohr-Rutherford theory says nothing about the nature of an atomic nucleus except that it is very small and very massive. The same questions that had arisen in the previous generation with regard to the atom now arise with regard to the nucleus—what is its structure, and how does it function? A few natural phenomena were known early in this century from which one could infer that things are going on inside a nucleus. The most remarkable of these was the existence of the different kinds of radioactivity. It seems that the nuclei of some atoms are not permanent entities but will, after a lapse of time, which may be as short as a fraction of a second or many times the presently accepted age of the universe, fire off particles and change into other kinds of nuclei. By measuring the energies of these particles and assuming that they bear somewhat the same relation to the structure of the nucleus as the light emitted from an atom does to the structure of the atom, it is possible to find out that nuclei also can have only certain definite energy levels and that the nuclear radiations are emitted when a nucleus shifts its energy from one level to another. Thus a new theory, something like the Bohr-Rutherford theory, is needed for a nucleus, but its mathematical details must be quite different, for here there is no single particle containing most of the mass of the system; rather, all the particles must be on an equal footing.

Two circumstances combine to make the nucleus a much more mysterious object than even the atom is. First is its small size—a typical nucleus has less than 10^{-4} times the diameter of the atom containing it. Second is the fact that nuclear particles are bound together much more tightly than the particles that constitute an atom. To disrupt a nucleus requires an energy a million times as great as that required to disrupt an atom, and for many years such energies were very difficult to produce in a laboratory. Now we know how to do it and can make energy-level diagrams for nuclei very similar to those shown earlier for hydrogen and helium. To study the sizes and shapes of nuclei is not enormously difficult; one merely performs more sophisticated versions of the experiment that originally established their existence. Rutherford deduced his results from collision experiments in which particles issuing from radioactive nuclei were directed at the nuclei to be studied. Nowadays one uses particles from special machines, which have been designed to produce them at any energy desired, and measures carefully the patterns formed by these particles as they speed into a liquid or solid target, collide with the nuclei, and bounce off into the empty space outside. From data of this kind, it is possible to determine that nuclei are exceedingly dense, that the particles in them are tightly bound together in a shape which is roughly spherical but which may depart from sphericity by being either longer like a football or flatter like a doorknob, and that the density of the electric charge throughout the nucleus is quite uniform up to the surface layer, where it falls off quickly. In Chapter 7 we shall discuss some of the models that have been made in an effort to understand nuclear structure.

CHAPTER *3*
Fields

Anyone who has ever played with a magnet has been in the presence of something that is very hard to explain mechanically. A magnet reaches out over apparently empty space and attracts a nail, but we have no intuitive understanding of how the force is exerted. Figure 3–1 shows one way of visualizing what is going on in the empty space. Physicists have known for a long time how to make pictures like this by sprinkling iron filings on a piece of paper so that they arrange themselves into little trains. Such a picture (and a very similar picture can be made of the space surrounding an electrically charged object) suggests that there is some sort of activity around a magnet that is caused by its presence. This activity—I purposely use a vague word in order to avoid any intuitive mechanical notions—is known as a field.

Let us take the word field as thus vaguely defined and build up its meaning as we go along. The magnetic field is what is in the neighborhood of a magnet that allows one magnet to interact with another. (The iron filings are really little magnets.) An electric field is what is in the space around an electric charge that allows one charge to interact with another. If you *hold* an electric charge near a magnet, nothing much happens; however, it is

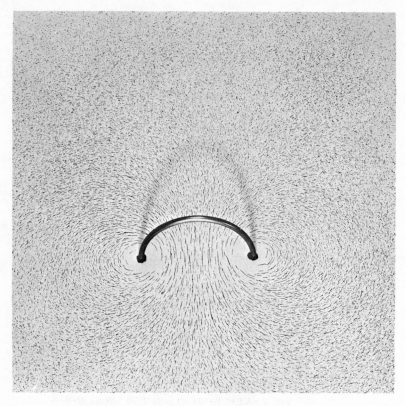

FIGURE 3–1   Pattern of iron filings around a magnet.
(From Physical Science Study Commit-
tee, *Physics,* Heath, Boston, 1960.)

clear that electricity and magnetism are somehow closely related from the fact that if you *move* an electric charge near a magnet, a force is exerted between them, so that in some way charges, magnets, and motion are linked up into a single system.

Now, to minds of a certain philosophic bent, the idea that a force can be exerted between two objects over an intervening distance with nothing at all in between to transmit the force seems to be very distasteful. This may be seen in the following paragraph of a letter from Newton to Richard Bentley, a prominent theologian who was writing a set of sermons designed to prove the existence of God from Newton's physics:

> That Gravity should be innate, inherent and essential to Matter, so that one Body may act upon another at a Distance thro' a *Vacuum*, without the Mediation of any thing else, by and through which their Action and Force may be conveyed from one to another, is to me so great an Absurdity, that I believe no Man who has in philosophical Matters a competent Faculty of thinking, can ever fall into it. Gravity must be caused by an agent acting according to certain Laws; but whether this Agent be material or immaterial, I have left to the Consideration of my Readers.

We are going to see in a moment a good reason for not believing in "action at a distance," as it came to be called; let us, just briefly, look at the alternative. We may suppose that there is some material medium in the space between magnets which is responsible for the force in the sense that magnet *A* acts on the medium and the medium acts on magnet *B*, so that it is almost as simple as though one magnet were pulling the other by means of a string. This medium has always been called the ether. The eighteenth-century philosophers were greatly concerned with the properties of the ether, and almost everyone (Newton was an exception) believed that these properties would ultimately be explained, like everything else, by Newtonian mechanics. It was later feared that there were several different ethers—one for electricity, one for magnetism, one for gravity, and one for light.

Light was at first thought to be some sort of pressure or succession of brief pulses transmitted by the ether; later, the suggestion that it was a regular wave traveling through the ether gained ground and by 1820 was almost gospel. Newton had believed that light consisted of particles, because of the sharpness of optical shadows—sound, for example, which is known to be a wave, travels around corners. The wave or pulse theory of light as it then existed could not answer his objection. Newton himself had performed experiments that seemed to require light to have a definite wavelength, but these he proposed to explain by endowing his particles with mysterious "fits of difficult and easy reflection." And he had also a profound inner sense of the uniformity of nature. He was a thoroughgoing atomist, as appears from his writings, and the idea that light was itself an atomic phenomenon was too attractive to dismiss as long as it was remotely tenable.

The experiments suggesting light to be a wave strengthened the position of those who postulated the existence of the ether, for if one thinks in mechanical terms, a wave must be a wave in some medium, as ocean waves are waves on the surface of water, and sound waves are waves of pressure in the air. It was difficult to build a mechanical theory around the ether because such a medium must have a number of conflicting properties. For example, the planets must be able to move through it freely, which means that it must be exceedingly thin; on the other hand, light must be able to travel through it at great velocities, which means that it must be highly incompressible. That the luminiferous ether could be both thin and incompressible at the same time was not easy to imagine. To invent ether that would transmit the effects of electricity, magnetism, and gravitation was hard enough, but the task was made much harder by the fact that electricity and magnetism are closely related—a moving electric charge produces magnetism and is acted upon by it. Much of the nineteenth century went past while the idea needed to unlock this complex was being sought, and of course, when it was found, it was not recognized by those who needed it most.

LIGHT AS A FIELD

Leaving out such questions as how matter can move freely through a fluid ether, in 1861 an ether theory of electricity and magnetism did appear. This was the achievement of James Clerk Maxwell, a British physicist who painfully put together a model of the ether that would explain all the phenomena of electricity and magnetism then known. He first summarized these phenomena in a set of differential equations, which are neither very numerous nor very complicated and which are today still regarded as essentially correct and complete. He then invented a highly complicated ether that would explain these equations in terms of various swirling motions, currents, and stresses within it. Out of this impressive theory came a dividend: the differential equations have solutions in the form of a wave, whose properties can be calculated in detail by purely mathematical means. When the wave's speed is calculated, it turns out to agree very closely with the speed of light—about 186,300 miles per second.

Maxwell correctly concluded that he had stumbled onto the theory of light. His theory explains the emission of light as an effect of moving electric charges and its propagation through space as an effect of the interplay between electric and magnetic fields. But it will be noticed that this explanation does not depend upon the ether; it depends upon Maxwell's equations. In the next three years the theory underwent a remarkable transformation, and when Maxwell presented it to the Royal Society at the end of 1864, only the introductory section of his paper mentioned the ether at all. He had, in effect, constructed a great edifice and then removed the scaffolding—some people would have said the foundations. The main part of the paper consisted only of the equations and deductions to be drawn from them—the ether had disappeared as unnecessary and irrelevant because Maxwell had begun to realize that everything that was verifiable in his theory resided in the equations themselves. The ether did not manifest itself in the equations except insofar as it gave them a mechanical explanation. Light, in Maxwell's theory, is a wave, and it is a wave in a field that is both electric and magnetic—in short, an *electromagnetic* wave field. This wave field can be described mathematically, but the sense of the description is hard to convey other than mathematically because it cannot be related to anything in ordinary experience. It has, and needs, no mechanical explanation that we know of, and any attempt to conclude that because light is a wave it has to be a wave in some medium seems to lead to irrelevant notions that have no experimental consequences.

In the following few years, Maxwell's theory was received with that dazzling indifference which has so often greeted entirely new contributions to science. It was not altogether ignored, however, and in the 1880's a German experimental genius named Heinrich Hertz set himself the task of verifying by experiment at least some of the theory's consequences. It is difficult to create light except by hitting atoms, and what Hertz wished to do was to create a wave of Maxwell's kind in such a way that one could see exactly how it was being created. The trouble with light is that the frequencies are too high to be readily observed in the laboratory—the electric charge that moves in order to produce light must oscillate about 10^{15} times per second—and so Hertz proposed to study a radiation analogous to light but at a very much lower frequency. Hertz's waves would be many octaves below the visible range of frequencies, bearing the same relation to light waves as low-pitched sounds do to high-pitched ones. To generate these waves, Hertz created electric circuits in which charges moved rapidly back and forth, and he was able to show that a wave is indeed launched out into space from such a circuit. When the wave encounters another similar electric circuit, it causes another current to flow, and by means of this second current the wave can be detected. It seems extraordinary that a discovery with such obvious possibilities in the communications field was ignored for over a decade, but nobody realized that radio waves, as they are now called, could be amplified so that a weak signal could be detected and made audible. This application was reserved for Marconi, Popov, and their followers in our own century.

LIGHT AS PARTICLES

Hertz's work verified Maxwell's equations, although it did not entirely stop the search for an ether theory. And as a purely incidental by-product, it yielded a new scientific discovery destined to have great importance. Hertz noticed that light falling on a metallic surface caused electric charges to be emitted from it. If one pumps away the air from around the metal surface and provides light in a suitable range of colors, electrons are liberated from the metal surface as soon as the light falls on it. This is called the photoelectric effect. Now, one explanation of this phenomenon is that the light gradually delivers energy to the metal, and that the electrons slowly start to shake as the energy builds up; they vibrate harder and harder and eventually shake themselves loose and leave the surface. It is easy to calculate how long it takes to build up the amount of energy the released electrons have—it turns out to be of the order of a billion years. And yet the photoelectric effect occurs instantaneously, even in very weak light. One is led almost inevitably to conclude that there is some sort of impact taking place at the metal surface, although this implies an atomic or particle theory of light rather than the wave theory that prompted the experiment. The thing producing the impact is called a photon. A very important discovery of the early 1920's was the Compton effect, which is a simple impact between one photon and one electron that appear to collide (as shown in Figure 3–2) according to the laws of conservation of energy and momentum, just like two billiard balls. The spectacular discovery of the double aspect of light—that it has some properties of a wave field and some of a particle—was largely the achievement of Albert Einstein in 1905 and thereafter. It set the dominating tone of thought in physics in the first half of this century, and a great deal of effort has gone into clarifying the resulting situation.

I must emphasize again that in considering the behavior of light one must free himself as much as possible from the mechanistic picture. It is clear that a mechanical wave cannot also act like a particle. Everyone agrees to this fact, but there are two common reactions to it. The first is to say, "All right, then the whole business cannot be understood." The second is more reasonable: to recognize that the wave and particle are not mechanical and that our instincts concerning the behavior of waves and particles, developed largely from familiar experiences, are not reliable in explaining to us how this particular kind of field is going to function. We are here involved in the construction of models, similar to those we

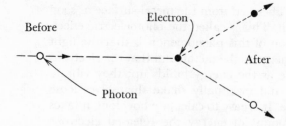

FIGURE 3-2   Mechanical, and therefore somewhat misleading, view of a Compton collision. A particle of light (photon) hits an electron initially at rest. After the impact, the electron moves off and can be detected. The photon also moves off. As it has given the electron some of its energy, a change in its wavelength has occurred, which can be measured. The measured change agrees with the result of a simple calculation.

have discussed earlier, designed to emphasize certain aspects of a situation in presenting it to our minds while playing down others. The unusual feature in this instance is that the two models, both useful, seem to have nothing to do with each other. They are bridged only by a numerical relation, first given in 1905 by Einstein, which states that the energy of a photon is proportional to the frequency of the associated wave: $E = hf$, where E is the energy, f the frequency, and h the constant of proportionality, known as Planck's constant.

Even if one does not take the models seriously, it is something of a puzzle that light should have this dual aspect or, to put it another way, that what appear to be particles of light should differ from other material particles in having a wave field associated with them. It was reserved for a young French student named Louis de Broglie to set the matter straight. He did this in 1923 by making what is probably the most fruitful abstract assumption that anybody has made in twentieth-century physics: namely, that if light is a wave field with aspects of a particle, then the uniformity of nature suggests that particles, that is, matter, must therefore have aspects of a wave field. De Broglie further proposed certain experiments to demonstrate the wave nature of electrons. These experiments are not particularly difficult to perform, and within the next few years evidence was accumulated in this country and in England to show, as in Figures 3–3 and 3–4, that elec-

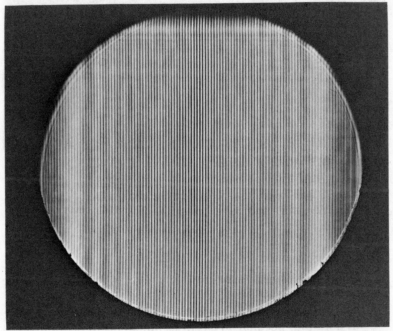

FIGURE 3–3   Pattern of interference produced by a Fresnel biprism, which divides a beam of light and then recombines the two parts. Light bands appear where the two waves reinforce each other, and dark bands where they cancel each other out. (From R. W. Ditchburn, *Light*, Blackie, Glasgow, 1952.)

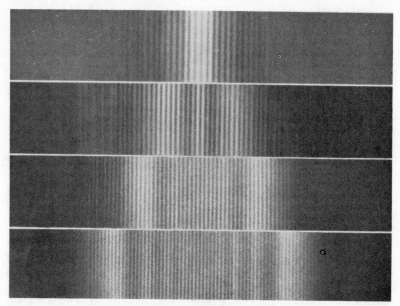

FIGURE 3–4 Pattern of interference between the two
parts of a divided beam of electrons,
produced in an arrangement analogous
to the Fresnel biprism. (From G. Möl-
lenstedt and H. Düker, *Z. Physik,* **145,**
377 [1956].)

trons have a wave nature closely analogous to the wave
nature of light.

The property of acting both like a particle and like a wave
is not at all peculiar to electrons but seems to be a universal
property of all matter. The wavelength of light is already
short; that of electrons is shorter still, and as we go up the
scale of masses from electrons to protons to whole atoms to
golf balls to planets, the wavelength gets shorter and shorter;
when a piece of matter is bigger than a few atoms, the wavelength is so very short that we have no way of investigating it
experimentally at all.

In order to measure the length of any wave, we must have

some sort of mechanical system that differentiates between waves of different sizes—a mechanical system with a periodic structure. To measure the wavelength of light, we rule fine lines, 20,000 or so per inch, on the surface of a piece of glass, and the light that passes through or reflects off it is thereby broken up according to its color. (It is perhaps a surprise to learn that a peacock's feathers contain no blue pigment, but only a regular scaly pattern adjusted to blue light.) For shorter wavelengths the lines must be closer together. Since there are limits to the fineness of ruled gratings, very short wavelengths are measured on a crystal, in which parallel layers of atoms replace the lines on the glass. Crystals are suitable for determining the wavelengths of electrons of very moderate energy and neutrons and protons of exceedingly low energy. It is just possible to measure the wavelengths of helium atoms and hydrogen molecules. There is no ultimate physical reason why some new technique may not be discovered with which the shorter waves of matter might be investigated, but no one has yet thought of it.

We must add to Maxwell's electromagnetic field several more fields—let us call them matter fields—corresponding to distinct entities such as electrons, neutrons, and protons and various composite fields representing atoms and molecules. The most fundamental level of physics today can be called field physics, and its program is to develop the experimental technique and physical insight necessary to understand the various fields, their structure, and in particular their interrelations. They correspond rather closely to the ethers of the nineteenth century, except that they are not thought to obey Newtonian laws. People who inaccurately picture science as a succession of towering structures of hypothesis that one after another come crashing down often use the ether theory as an example. It is a very bad one, for what it really illustrates is the true development of science—a process of gradual refinement.

It is fashionable to refer to Newton's corpuscular theory of light as one of his big failures. Although pointing out the errors of famous people is fun, I would suggest that this theory was instead one of his greatest achievements. It is true that he got the details of it all wrong. It is also true that, having been published prematurely, Newton's theory of light did more harm than good and hindered progress for the greater part of a century; but for this the man is less to blame than his reputation, and the feeling for the profound inner consistency of nature that was shown in Newton's theory is just the same as that which prompted de Broglie to his brilliant discovery. We must conclude that in the longest historical view Newton was right, as he was so often in his career, and for the right reason.

MATTER AS A FIELD

It was all very well for de Broglie to find that the elementary particles of matter have a wave nature as well as a particle nature, but the most perplexing issue in physics in the years preceding this discovery was to understand atomic structure. (It will be remembered that the Bohr theory seemed to have run into a dead end at the hydrogen atom and could not be pushed any further.) Atomic structure poses problems that can be resolved only by computation. One has to make a particular model and then calculate detailed properties of systems composed of large numbers of atoms. These properties are then compared with those found by experiment, and if the answer comes out right, the conclusion is that the model is a good one. We have already seen how this procedure worked for Niels Bohr with the hydrogen atom.

In order to see whether de Broglie's matter fields could be used to explain the behavior of an atom, it was necessary to have an exact mathematical statement of their properties. Historically the position was exactly the same as that confronting the electromagnetic theory before Maxwell appeared, when there were many facts known about light, together with a general idea of radiation moving through space, but no equations and little that could be calculated. The essential step for the resolution of atomic structure was taken in 1926 by the Austrian physicist Erwin Schrödinger, who wrote down a wave equation for matter that is analogous to Maxwell's equations except that it is a little simpler. When an equation has been proposed, we begin to move toward precision of thought. The new equation must be solved in mathematical terms, and when compared with experiment, if the calculation and the data are good enough, it must give either a right answer or a wrong one. Schrödinger's equation began at once by giving the right answer for the spectrum of hydrogen. It was more difficult to understand the spectrum of helium, but soon approximate methods were developed, and after a year or so the right answer for it was available.

This put the Bohr theory in a rather curious light, because the idea of an electron's wave field in some way mixed up with the proton's wave field to produce a structure having certain well-defined energy levels and emitting certain well-defined wavelengths of light seems pretty remote from the planetary model of Rutherford and Bohr. In fact, what is surprising is not that the Bohr theory failed to explain the spectrum of helium but that it ever gave the right answer for that of hydrogen. We must regard this as a coincidence—fortunate because it encouraged the investigation of atomic structure in terms of a mechanical system with definite energy levels but unfortunate because it was somewhat misleading to physicists of the time.

It now turns out that the Schrödinger theory, which is known as quantum mechanics, will, if one works hard enough at the mathematics involved, explain anything one wants it to about the chemical and physical structure of matter above the nuclear level of smallness. This is not to say that everything has been explained, because often one encounters difficulties of mathematical technique that prevent calculations from being carried out with a finite amount of effort and, worse still, shortcomings in imagination that do not allow one to think of the right things to calculate or the right approximations to make. However, it is true that nothing that has ever been observed in the laboratory suggests any fault in the picture of reality offered by quantum mechanics. I do not mean to imply that anyone thinks of quantum mechanics as the ultimate physical theory. In Newton's time, after all, there were no facts known in the atomic or relativistic domain to challenge the correctness of his laws. It is not even safe to claim that the ultimate physical reality will always be thought of as a field. All that can be claimed is that ground has been won against the unknown that need never be given back.

To live in a world composed of fields seems to be a very different thing from living in a world that operates like a mechanism. Most of our language and our philosophic viewpoints are still based on a mechanistic conception of the universe, to the extent that they are based on anything scientific at all. There is no need here to go into the profound modifications that quantum mechanics requires of our view of reality, but because it is much spoken of, one new idea that has emerged should be mentioned: this is the principle of indeterminacy or uncertainty.

Clearly one of the principal differences between the idea of a particle and the idea of a wave is that a particle can be localized in space and a wave cannot. To be sure, if one imagines the wave corresponding to a directed flash of light, as in Figure 3–5, the light will be relatively localized in space in the sense that there is no light before the flash and none after it. But still the light does not arrive at the eye at one distinct moment in time the way a bullet arrives at its target.

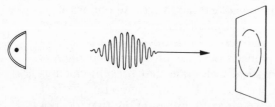

FIGURE 3–5   A flash of light emitted from a spark in a parabolic reflector moves as a localized and slowly spreading wave. After a short time interval, it arrives at its target to cover a certain surface area.

It would appear at first sight, therefore, as though quantum mechanics and its wave equations were inadequate to describe the arrival of a particle. Common sense says that it must be possible to localize a particle at a certain point in space at a certain instant, whereas a wave can only be confined to a fairly narrow region.

In 1927 Werner Heisenberg attacked this apparent inadequacy in a daring way. He said that the fuzziness which is an essential property of waves in the wave theory, far from being a disadvantage in the theory, is the exact reflection of a universal property of nature that had not previously been noticed—that it is actually impossible to localize a particle with infinite precision by any experimental procedure whatever. The exact statement of Heisenberg's indeterminacy principle involves considerations that we need not go into here, but it is important because it asserts that our ability to know things has absolute limits that seem to correspond exactly to limits inherent in quantum mechanics. This correspondence is one of the features of quantum mechanics that now give us confidence that it mirrors nature accurately.

One of the peculiar features of Schrödinger's wave equation is that it does not satisfy the requirements placed upon physical theory by the theory of relativity. Normally, this would not be considered very important, since the electrons in an atom are not moving at extremely high velocities, and it is known that relativistic considerations normally come into play only when velocities approach the speed of light. Still, the equations of quantum mechanics are supposed to be getting at something very profound, and relativity is also very profound; thus it is natural to expect that some type of field theory exists that takes the relativistic mathematical form which Einstein had shown that all physical theories have to take. The relativistic form of Schrödinger's equation was first given by an English research student named Paul Dirac in 1927. It differed from Schrödinger's equation rather markedly, but Dirac was able to show that when it was applied to such problems as finding the energy levels of hydrogen atoms it gave almost exactly the same answers as Schrödinger's equation and that the very small differences corresponded with differences encountered in nature.

The most significant result of Dirac's theory, however, was the discovery that the spin of a particle is a necessary and inherent feature of the particle itself. We have mentioned the property of spin in Chapter 2. No one has ever seen a spinning particle, of course, but its spin gives it special dynamic properties like those of a gyroscope or a top, which show up quite markedly in its experimental behavior. The existence of spin was first deduced from atomic spectra by two Dutch physicists, Samuel Goudsmit and George Uhlenbeck, in 1925. As with so many great discoveries, there were precursors: Compton, who did not get it quite right, and Ralph Kronig, a German student at Columbia, who let himself be intimidated by a more eminent man who thought he was crazy.

In order to include the spin of electrons in the first version of quantum theory, it had been necessary to put it in as a special *ad hoc* hypothesis to explain certain facts. Dirac showed, however, that his theory was not capable of describing an electron without a spin and that, furthermore, the amount of spin and magnetism that the electron has is exactly given as an automatic consequence of what might be called his relativization of Schrödinger's equation. Dirac's discovery was of great importance both because it filled in this gap and because it was a sign that physics was on the right track if two previously known deep physical ideas— those of the matter field and of relativity—could be brought together to yield something new.

QUANTUM MECHANICS

Obviously we cannot collect a mass of experimental information on the properties of light and electrons, establish that half of it can be explained by the particle model and half of it by the wave model, which has no recognizable point of contact with the particle model, and then pretend to ourselves that physics had done its job. If nothing else, we must produce either a new intuitive model that explains the paradox of the two others or else a mathematical substitute for it. It is the latter alternative that has been successful. Quantum mechanics does not arise out of a physical model but tries to reach the nature of reality through mathematics. With the passage of time, a remarkable thing has happened. To the physicists of Schrödinger's day, quantum mechanics was an abstract mathematical procedure that led to correct answers. But then a new generation began to grow up, who knew only by hearsay of the terrible struggles of imagination that their elders had gone through to invent and master this system of abstractions. To the young it was no such thing, but simply the way the world operates, an economical and natural description of physical processes that can be grasped and used intuitively as easily as an architect utilizes relations between forms and positions in space. Like architecture, quantum mechanics has developed a language for communicating rapidly and accurately without bringing in extraneous or ill-defined notions. It expresses itself in special new terms—*wave function, matrix element, transition probability, renormalization,* etc. —each of which carries the weight of physical reality, and it is in these terms that the next stage in our understanding will laboriously be worked out. They form in fact a new model. Like others it is rooted in experience, but it is different in that the experience is one of mathematical calculation. Although its sense cannot be communicated without mathematics, the following paragraphs convey something of the picture of reality that has come out of it.

The remarks that follow do not really constitute a "popularization" of the subject, since popularizations usually involve a certain change of viewpoint, in this case a reinterpretation of mathematical results in terms of models that, though they may be dangerous and misleading if pushed too far, have the merit of being easily visualized. The discussion here will be relentlessly abstract, and the reader is urged to banish from his thoughts the image of a little raspberry-shaped lump of nucleus surrounded by whirling colored streamers that is supposed to denote "atom" in the semipopular press.

In order to fix our ideas somewhat, let us begin with light.

The theoretical viewpoint must base itself on an immense number of experimental observations relating to light, most of which can be roughly summed up under these three headings:

1. Light originates in moving electric charges and is in turn absorbed by other moving electric charges.
2. Light exhibits all the properties of wave motion, particularly interference and diffraction.
3. When light exchanges energy with matter, it often does so in definite, discrete amounts, depending on the experimental situation.

In order to provide a compendious idea of these and a few other facts, physicists represent light as a *quantized field*— a mathematical structure that takes on a definite value at every point of space at every instant of time. Even in a perfectly dark room, the light field is specified in a definite mathematical way through the fact that in interacting with matter it cannot give but can only receive energy. Moreover, the motion of this field is a wave motion according to *2* and is described by certain definite equations. (A mechanical analogue of this concept of field would be the surface of a body of water, which, if the water is disturbed, will vary in a pattern of ripples.) According to *1*, the only things that can disturb or, more exactly, interact with the field are moving electric charges. It is important that this coupling appears to be local; that is, the coupling of charges to a field at a certain point depends only on the value of the field at that point and not on its value at neighboring points.

The further characterization of this field as quantized takes account of such phenomena as the Compton effect and the photoelectric effect, which are grouped under *3*. When the field changes its state of motion (roughly, the amplitude and pattern of its waves) by an exchange of energy, it does so discontinuously, and the bits of energy exchanged are called *quanta*. Quantized behavior, which is the actual behavior on the microscopic scale, may be visualized in the following way. Imagine the field coupled to matter to be like a water surface with corks floating on it. If the surface is disturbed by ripples, all the corks will bob up and down. Now imagine that all the corks remain at rest in the moving water and then that one of them suddenly flips several inches into the air. This is the peculiar situation that exists when the field is quantized. Characteristically, the quantum field theory does not state which cork will move but only states the probability that any given cork will move within any given time interval.

It is not hard to build this discreteness into the theory, but when one does so, there are a number of consequences that seem to be necessary though they are perhaps unexpected. The most striking of these consequences is a weakening of the analogy of the liquid surface that was just used, for it is no longer possible to consider a field completely at rest. Even the field characterizing a perfectly dark room turns out to be full of fluctuating activity—it is continually exercising in a random way every possible mode of motion available to it (and they seem to be infinitely numerous). Nor are these peculiarities purely hypothetical, for where the field is coupled to matter it imparts to it something of these same fluctuations. The fluctuations are on a very small scale and are to be distinguished from phenomena like the Brownian motion in that they would exist even at a temperature of absolute zero. They have been studied carefully in the last twenty years, largely through their slight but perceptible effects on the spectra of atoms.

Now let us look at electrons. The word "electron" is perhaps misleading, but there is no better alternative. To the uninitiated, it suggests a little particle. To the expert, it denotes only a state of excitation of a whole quantized field. With this warning, we shall consider electrons as typical of electrically charged matter from the same standpoint as was adopted for light, noting the similarities and differences. Under headings corresponding to those previously given for light, we have:

 1. Electrons are rarely created or destroyed. Unlike the light around us, most of which lasts for only a few minutes (the time required for it to reach us and be absorbed after its birth in the sun), most electrons are about as old as the universe. The creation of an electron is a rather special event, occurring from time to time in certain naturally rare nuclei or in the interaction of cosmic rays with our atmosphere. (And when it does occur, there are compensating changes in other particles so that no electric change is created or destroyed.)

2. Electrons exhibit all the properties of a wave motion, particularly interference and diffraction.

3. When electrons exchange energy with light or other matter, they often do so in definite, discrete amounts, depending on the experimental situation.

4. The electron has a twin, the *positron,* identical with it except that the sign of its electric charge is reversed.

Except for *1* and *4*, these specifications are essentially the same as those for light. But *1* is only quantitative and not qualitative. As for *4*, since light has no electric charge, its twin would be an identical twin, absolutely indistinguishable and therefore not worth discussing. Wherein, then, do light and electrons differ? In several respects. First, they differ in the nature of the coupling between them; an electron can emit and absorb (that is, create and destroy) a single quantum of light, but a quantum of light cannot emit or absorb an electron. (The reason is simple: electric charge can be neither created nor destroyed. If, therefore, light creates an electron, it also creates a positron, so that the charges add up to zero.) Second, they differ in the nature of the motions of the two fields; both are wave motions, but they obey different sets of equations, just as water waves and sound waves have somewhat different mechanisms of generation and propagation.

One point should be noted with the greatest care: that the existence of electrons as particles is suggested by certain features of the field's interactions with other matter, but not at all by its configuration in space. Electrons are not to be regarded as lumps in the field—on the contrary, fields do not normally vary sharply in value from one point in space to another. If one speaks of detecting an electron at a certain point, he means simply that some sort of instrument placed at that point has clicked or flashed, registering a sudden quantized exchange of energy, and not that the field was momentarily particularly intense there. (Although we have talked only of electrons, the behavior of all the different fundamental kinds of matter is found to be much the same.)

One of the most striking properties of quantized fields is their behavior when two of them interact. Here the effect of the coupling is to produce a high degree of correlation between occurrences in the two fields. If, for instance, an electron is detected at a certain point, the light field at that point will be found to be in a state of turbulent excitation at the same time. The naive picture corresponding to this situation is a charged particle surrounded by a swarm of photons. One might guess from this that if the charged particle were suddenly accelerated, some of the accompanying photons might be flicked off, like water from a wet towel, and appear as real light. Mathematics supports this guess and predicts the wavelength and angular distribution of the light; experiment, in turn, verifies the predictions. This is typical of the way in which the particle picture can help one to understand and even to predict new phenomena. But it must be remembered that a picture such as this contains only as much truth as is given it by the underlying mathematical model and that its only function is to help bridge the gap between an abstract symbolic formalism and the set of intuitive concepts in terms of which we express our view of the world.

It is natural to ask, "How big is a particle?" Clearly, the quantum field theory does not concern itself with this question, since it does not speak of particles. But the question can be replaced by another, which for practical purposes is a satisfactory substitute: "Over how large a region does the correlation of intense fluctuations in two fields occur?" To this the theory gives (with some difficulty) a definite answer: Depending on the nature of the fields involved and on one's criterion of "intense," the size of the region is of the order of a few times 10^{-13} cm for most particles, though for the electron it seems much smaller. This is in agreement with the latest experimental results on the bombardment of nuclear particles by high-energy electrons. If one now considers a whole atom—a mixture of nuclear particles and electrons coupled together by electric forces—one can ask exactly the same question and find (rather easily, this time) an answer some hundred thousand times as great as the preceding one. This would be the size of an atom. We must remember that there is here no notion of little balls whizzing in colored loops around a central lump. That image is picturesque, and it is by no means useless in science, but it is of almost no help in understanding the modern theory.

The quantum field theory does have limitations. Those of which we are most conscious relate to our knowledge of how the various fields are coupled together. Although a great variety of calculations can be performed whose results agree closely with those of experiment, there are a number of details that are unsatisfactory from a mathematical point of view, and more than this, the theory withholds from us certain results that it ought to give. When light is produced in the motion of an electric charge, the theory predicts the mass (or, equivalently, the energy) of this light. But when an electron is born in a nuclear process, the theory does not predict what its mass will be. Instead, we have to put into the equations a number derived from measurement, and any other number would, from a mathematical standpoint, serve equally well. In fact, the most dissatisfying thing about the present theory is that it predicts almost nothing about the natures of the various sorts of field that are found to exist, and it is reasonable to judge that we are introducing field interactions into the theory in a way that is rough and empirical rather than exact and fundamental. The rich harvest of experimental data being gathered in many parts of the world will surely lead within the next few years to some idea of the nature of the relations between different types of matter and, after that, to a revision of some of our basic ideas about coupled fields. At present, we imagine all space to be filled by a superposition of fields, each named after an elementary particle—electrons, protons, various kinds of mesons, etc. As new species proliferate, it becomes more and more desirable that the future theory, if it resembles the present one at all, should contain but a single field, with the present types of matter corresponding to different modes of excitation of it. The main obstacle to progress seems to be that the information yielded by modern experiments is not of a kind that bears directly on this issue. We must, it appears, content ourselves with hints.

CHAPTER 4
Gravity

In the last chapter we saw that fields have two essentially different aspects, an aspect of force and an aspect of matter. The second is the more recently discovered, but the first is just as little understood. Of all the fields of force, gravity is the most familiar. It is by far the strongest of the various fields that we see acting in daily life, and we have a tendency to think of it as playing a dominating role in physics; indeed people thought of it this way up until the nineteenth century. But I have already mentioned that gravity is a weak force. The only reason that we experience so much of it is that we live in the neighborhood of a very large and heavy piece of rock, toward which everything is attracted. If, instead, we lived on an asteroid out in otherwise empty space, we would know very little about gravity, and it would be a laboratory curiosity like the electrical tricks that can be done with a comb and a piece of paper on a cold day. We know that our two hands attract each other by the force of gravity, but we can never feel so weak a force.

If we compare the strength of the gravitational field with that of the electric field in a situation where both of them are present, we get a surprising result. The force of electrostatic attraction between an electron and a proton (for example, in a hydrogen atom) is about 10^{28} times as great as the force of gravitational attraction between them. Gravity still has a special relationship to all the other fields of force, but it belongs at the end of the list. We have to think of it as being quite distinct from electric and magnetic forces because it is vastly weaker than they are.

We said in Chapter 1 that the gravitational field is crucial to Newton's theory of planetary motion. Newton's first law of motion states that an object with no force on it at all moves in a straight line at constant speed. The effect of the gravitational forces is to bend the planetary paths away from these straight lines into closed orbits. But a question arises: A straight line at constant speed with respect to what? From a practical point of view, one cannot describe any kind of motion without saying with respect to what the motion is taking place. Newton had an answer to this to which he was compelled by logic; it was that there is such a thing as absolute space. The properties of this absolute space were not very clear—for Newton they were at least partly theological—but absolute space did provide a standard of straightness by which a particle could regulate its course. Many philosophers in Newton's time, particularly Leibniz and Huygens, vigorously disputed Newton's contention. They said that geometry as we know it, which is the study of space, consists entirely of the spatial relationships between objects and that nothing in our experience requires the existence of absolute space.

Leibniz and Huygens were unable, however, to answer Newton's argument: Consider a pail of water hanging at the end of a long rope. If the rope is twisted and then allowed to untwist, the pail will spin and the water will climb part way up the sides of the pail, making the surface of water concave. If the vessel is not turning, the surface is essentially flat. Therefore, Newton argued, there is an absolute difference between a vessel of water (or anything else) that is rotating and the same vessel when it is not rotating. This absolute difference he referred to as absolute space, saying that the pail in the one case was rotating with respect to absolute space and in the other case was not.

A more quantitative example of the same argument is found if we consider the solar system. The planets, moving according to Newton's laws, travel in ellipses. The long axis of each of these ellipses is almost exactly fixed in space. (The extent to which it is not exactly fixed in space is also given almost exactly by Newton's laws and can be taken into account.) Let us imagine that the solar system is alone and that no other stars at all are visible to the naked eye. Suppose then that one wanted to know what absolute space was. By observing relative positions of the planets and finding out which way the axes of the various ellipses pointed, one could make a diagram in space with certain fixed directions, and these would be taken as directions in absolute space. Now let us suppose that the stars are lighted up and we see them for the first time. We would then be amazed to find that the set of lines in space that were deduced from a purely mathematical argument turned out to point always to the same stars in the sky, that is to say, as if Newton's absolute space were in some way attached to stars. This is exactly the situation that exists, and one has the unpleasant sensation of two scientific facts running along on independent parallel tracks because one cannot see the connection between them.

Suppose we think of a little particle moving through space without any force acting on it. It moves in a straight line at constant speed, but we no longer have to invoke absolute space in order to explain what we mean by this statement, for we may say just as well that the motion is in a straight line with respect to the fixed stars. (The fixed stars are not really as fixed as they look. All of them are moving to some extent in one direction or another, but the apparent motion is very slow and can be ignored or averaged out.)

This circumstance excited the interest of one of the great speculative physicists of the nineteenth century, Ernst Mach of Vienna. He reasoned that if a free particle moves in a straight line at constant speed with respect to the fixed stars, as is true from an observational point of view, and if any deviation from such a motion requires a force to be exerted, as is also true, then in a certain sense the fixed stars must cause the force. This idea is known as Mach's principle. Mach further proposed that if one were to imagine all the fixed stars taken away, leaving a single particle alone in the middle of a completely empty universe, there would be nothing in all of physics to tell us what laws of motion this particle would obey. He remarked that if the stars in some way make their presence felt by a particle moving among them, in the sense that the particle requires a force to make it depart from uniform motion with respect to them, then some sort of influence must reach from the stars to the particle. It is repugnant, and anti-Newtonian, to assume that one can have an action without an equal opposite reaction, and we might suppose that the particle itself reacts back on the stars and exerts a tiny force on them. This suggests the existence of a field of force in space, a force exerted by the stars on the traveling particle and by the traveling particle on the stars.

THE PRINCIPLE OF EQUIVALENCE

It occurred to Einstein in the years following the publication of the special theory of relativity in 1905 that Mach's field of force was probably very closely connected with gravity and might be either Newtonian gravity itself or some extension of it. We know, after all, that gravity is a universal property of all objects, as are mass and the tendency to move uniformly, and perhaps these properties are all related. In particular, if the gravitational influences of the stars have something to do with the inertial properties of a small mass, then it must also be true that the inertial properties of the mass have something to do with its own gravitational properties.

A remarkable fact about gravity, which falls in with this line of thought, is that if one drops a heavy weight and a light weight from the same height, they will both reach the ground at the same time. This is the experiment that somebody (not Galileo, but possibly Simon Stevin of Bruges) performed from a tower in Pisa. Of course, we all know they will not reach the ground at exactly the same time because air resistance differs on different objects. But there is good evidence that in the ab-

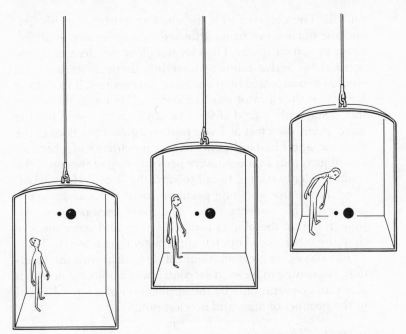

FIGURE 4–1   To an observer in an elevator accelerating upward in force-free space, two unlike balls appear to fall to the floor at exactly the same rate. To an outside observer, they do not move at all.

sence of air resistance the weights would fall at very nearly or exactly the same rate.

Einstein proposed to examine the consequences of assuming that the weights fall at exactly the same rate and to attempt a physical picture of gravity with this fact in the foreground. He imagined an elevator in the middle of space, with no appreciable gravitation acting upon it. It has a string attached to the top of it, and there is some agent by which the string can be pulled. In the elevator are an observer standing on the floor and two different-sized masses floating freely near the ceiling (Figure 4–1). What happens if the string is pulled? The observer feels the elevator pushing on his feet, and the masses, not being attached to the elevator at all, remain at rest in space. The elevator floor accelerates up toward them and a moment later hits them. Note how this would be interpreted by the man in the elevator, if he did not know that the elevator was free to move. He might report like this: "Suddenly a field of force began to act. I could feel the force on my own feet as I was pushed against the floor of the elevator, and I had independent ocular evidence of it because two objects that I thought were perfectly safe at the top of the room suddenly started to fall toward the floor." He is asked, "Did you notice anything peculiar about the way these objects fell?" He answers, "One thing was very peculiar. Although one of the objects is much larger and more massive than the other, they both fell at precisely the same rate."

The physics of Newton is quite able to deal with this situation. According to Newton's second law, an object of mass m is given an acceleration a by the application of a force F equal to the product of mass and acceleration,

$$F = ma.$$

If all objects fall with the same acceleration, then a is a constant quantity in this relationship, and the force is proportional to the mass. But we have already seen that this proportionality of the force acting upon a body to the mass of the body is an essential feature of Newton's theory of gravity; we have thus come back to that famous hypothesis, and in fact by an argument quite similar to that used by Newton in the first place. Nevertheless, it is valuable to look at a gravitational field in such a way that this property is built into it at the very beginning and is essential to the description, quite apart from any of the algebra the theory may contain.

Einstein now went further. He asked, "What if *every* experiment that the man is able to do with moving weights in his little accelerating laboratory were to give exactly the same results as if it were done in a laboratory at rest on the surface of the earth and subject to its gravity?" If this were so, then one would be able to say that the force of inertia, as encountered in Newton's second law of motion, is the same as a gravitational force. The hypothesis that gravitational and inertial forces are basically equivalent is known as the *principle of equivalence*. Several experiments can at least be imagined with a view to testing its truth. In the first place, suppose that we have a set of objects—a piece of iron, a book, a stone, and an onion, all of exactly the same weight. (Remember that the weight of a thing is the force with which gravity pulls it toward the earth.) Then we measure their inertial masses as defined by Newton's second law, by seeing whether a given force will produce the same acceleration in all of them. The question is, will the inertial masses be exactly the same? Or to put it more succinctly, is weight always precisely proportional to inertial mass? If there is a little difference from one object to another, then the principle of equivalence cannot possibly hold.

In the late seventeenth century, Newton carried out experiments with pendulums of wood and metal in his chambers at Trinity College in Cambridge, which allowed him to conclude that mass and weight were quite accurately proportional to each other. In the early years of this century, Baron Roland Eötvös of Hungary carried out essentially the same experiments, but to an accuracy of one part in 10^7, and recent experiments conducted by Professor Robert H. Dicke at Princeton University have increased the precision to one part in 10^{10}. So far, the results support Newton and Einstein.

The principle of equivalence has been introduced here as an alternative way of explaining phenomena that are already clearly explained by Newton's theory of gravity. But the test of a scientific idea is to see whether it leads to anything new, and to this end we look for some phenomenon that falls outside the domain of mechanics governed by Newton's theory. Let us, for example, think about the effect of gravity on a beam of light. To get a notion of what it might be, let us consider Einstein's famous elevator, with a beam of light projected across from one side to the other, forming a spot on the opposite wall (Figure 4–2). Now suppose that a force is applied to the string. Light is emitted from the light source and starts across the elevator. While it is in motion, the elevator

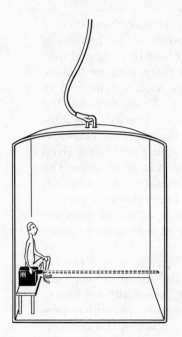

FIGURE 4–2　A beam of light is directed across an elevator at rest. Then, while the light is in transit, the elevator rises a little, and the light hits a lower spot on the wall than it did before.

picks up speed, and the light reaches the opposite wall of the elevator farther down than it would if the elevator were not accelerating. An observer in the elevator would deduce that the same downward force that affects everything else in the elevator is also bending the light downward. One would, therefore, conclude that light has weight and that it is affected by the "gravitational" field in the usual way. Now let us estimate how big this effect would be.

The normal rate of acceleration of objects falling to the earth is about 32 feet per second in each second. The apparent deflection of the beam, given by x in the figure, is the distance the elevator rises in the time available. Since light travels at about 10^9 feet per second, it takes about $10/10^9$, or 10^{-8}, second to cross the elevator. During this time, the elevator's speed increases by 32×10^{-8}, or 3.2×10^{-7}, feet per second, and its average increase in velocity is half this, or 1.6×10^{-7} feet per second. During the interval of 10^{-8} second, it therefore rises $1.6 \times 10^{-7} \times 10^{-8}$, or 1.6×10^{-15}, feet more than it would have if it were unaccelerated—a little less than the diameter of an atomic nucleus. The measurement would be impossible to make.

We accordingly must look for some other property of light to use. A suitable one is the light's frequency. Imagine that the source of light is located at the top of the elevator and shining toward a receiver at the floor (Figure 4–3). At a certain moment a little pulse of light is emitted from the ceiling and starts toward the floor. Then, as the figure shows, the elevator accelerates, the floor moves up toward the ceiling, and by the time the light arrives at the floor, the floor is moving upward toward the source with a velocity that it did not have at the time the light was emitted. Effectively, therefore, the receiver of light is moving with respect to the source. We now need a way of detecting this fact.

The Doppler effect, quite a familiar phenomenon, does precisely this. Everybody is familiar with the effect of riding on a train past a ringing signal bell and hearing the pitch drop as one passes it; an automobile with its horn blowing as it passes us on a road gives us the same effect. If the source of the sound is moving with respect to the receiver (it does not matter which is moving), then there is a change of frequency at the receiver. Now let us estimate the change of frequency in this case. Suppose that the room is a generous 100 feet high. Traveling at 10^9 feet per second, a flash of light would take $100/10^9$, or 10^{-7}, second to travel from the ceiling to the floor. Accelerating at 32 feet per second each second, the floor would acquire a speed of 32×10^{-7}, or 3.2×10^{-6}, feet per second by the time the flash arrived there. The rule governing the Doppler effect is that the fractional change in the light's frequency or its wavelength is the velocity of the source or receiver divided by the velocity of light. This frac-

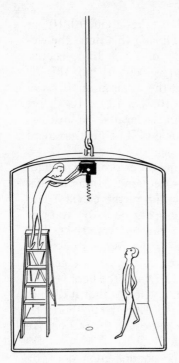

FIGURE 4–3 When a pulse of light arrives at the floor
of an accelerating elevator, the floor has
acquired an upward velocity with respect to the velocity of the light source
when the light was emitted.

tional change is $3.2 \times 10^{-6}/10^9 \cong 3.2 \times 10^{-15}$, or about
three parts in a million billion. But although this fraction is
almost unimaginably small (it corresponds to approximately
10^{-5} inch in the distance from the earth to the moon), there is
no essential physical principle that makes it undetectable. As
a matter of fact, we shall see in the next chapter that it is just
barely possible to pick it up.

Here is a new fact whose existence is predicted by the principle of equivalence. Unfortunately, it can also be explained
from the mechanical viewpoint if we assume that light is

mechanical and allow ourselves a few more liberties. If the experiment works, we can say that it is in accordance with the principle of equivalence but not that it proves the principle to be correct. On the other hand, if the experiment does not come out as expected, the principle must be wrong.

INERTIA

Now let us return to Mach's principle—that inertial forces in some way reflect a field property of all the mass in the universe—and consider its present status. The principle is very possibly true. We do not know because we do not know enough about how the universe is put together. Einstein was never able to form a complete theory corresponding to Mach's ideas, and in order to do this properly, one may have to know all about the large-scale structure of the universe.

The expression "all about the . . . universe" refers to its past history as well as its present condition. Physical influences seem never to act instantaneously. The inertial properties of a particle are not really determined by the present positions of all the stars but by their positions in the past when their supposed influence started. According to Einstein's theory, which is not necessarily true, the influence would travel at the speed of light—so fast that over all ordinary distances the delay would not matter at all. But some of the objects of the universe seen through a telescope are so distant that the light started from them billions of years ago, and therefore what is seen through the telescope is not the present state of the universe at all but its state at progressively earlier epochs as one looks farther out. However, this situation may be just what is appropriate to the question at hand, because the influence that produces inertia may be traveling at the same speed. The question is complicated, and there is no very simple answer until we know very much more than we do now.

We must also admit that we are not entirely sure what field we are talking about. Still, there is one property that seems to characterize all fields and, in addition, seems eminently reasonable: the farther one is from the object that is creating the field, the weaker the field is. The gravitational field falls off inversely as the square of the distance from the object that causes it, and so does the electric field; the magnetic field is somewhat more complex, but it too becomes weaker as one goes farther away. If this property applies to the field responsible for inertial effects, then we might expect that a distribution of mass relatively near the place where an inertial experiment is being carried out would have more effect on the result of this experiment than a similar distribution of mass at a great distance away and that, if there is a large distribution of mass somewhere in the vicinity of the experiment, the direction toward this mass could somehow be distinguished in its inertial properties from any direction perpendicular to it.

Living on the earth as we do, we are about halfway out toward the edge of a galaxy consisting of some billions of stars. To one side of us is the galactic center with its tremendous mass. In other directions space is comparatively empty. There are of course a vast number of other galaxies surrounding ours, but they are all relatively far away. Let us suppose, therefore, that the effect of the galactic center on one side of us is larger in producing inertia than the effect of similar concentrations of mass at greater distances. Then is it possible that this center could produce some special effect?

In Newton's second law, the mass m has no directional properties at all, and the law can be taken to imply that it takes exactly as much force to give the same object a given acceleration toward the east as toward the north. The question we ask is, is this statement exactly true, or is it only very nearly true? (There is no question of its being seriously inaccurate; if that were so, the inaccuracy would have been discovered by astronomers long ago.) What interests us here is the possibility of a very slight anisotropy (that is, directional character) of inertia that could be detected by delicate measurements. Let us suppose, for example, that a pendulum is swinging above the surface of the earth at the latitude of New York City in a north-south direction. If we note its period of swing and then wait for some 12 hours, the earth will have turned halfway around, and we will easily see that the pendulum is swinging along a different line, roughly 90 degrees away from the first. Presumably the force of the earth's gravity on the pendulum will be unchanged. On the other hand, the east-west line, if previously it pointed toward the galactic center, will now point about 90 degrees away. If the inertial mass were, for example, a little greater as a result of this, then we would expect the pendulum to swing a little more slowly. It is very difficult to perform the experiment with a pendulum because it must be done with great accuracy (and besides, a pendulum works by gravity, which we do not fully understand), but we shall see in the next chapter that there is a kind of oscillation inside an atomic nucleus, which has the directional quality this experiment requires and which can be used in a very sensitive test of the isotropy of inertia.

STRONG FIELDS

The effects we have been discussing are all very small; if they can be detected experimentally at all, it is only with great skill and hard work. For this reason we really know very little about gravity. Our ignorance is especially frustrating because we have the detailed theory of gravitation in Einstein's last masterpiece, the general theory of relativity, announced in 1915 but never yet subjected to a really searching experimental test. Gravity in Einstein's theory is slightly different from what it is in Newton's and the small amount of experimental evidence favors Einstein. Most notably, there is a slight drifting of the orbit of the planet Mercury, for which Newton's theory cannot account but which Einstein's theory seems to explain exactly. Einstein's equations tell us of astonishing things that would happen in gravitational fields billions of times as strong as those we know, and for half a century it has been a matter of regret that all the fields encountered in nature are so weak.

It seems that the regret may have been premature. Early in 1963 Maartin Schmidt of the California Institute of Technology, acting at the suggestion of Cambridge University astronomer Fred Hoyle, photographed a faintly luminous stellar object known as 3C273 and discovered, by astronomers' special techniques, that it is over a billion light-years away. We now know several things about 3C273. It is enormously bright, approximately 100 times as bright as a normal galaxy consisting of billions of stars, although its size cannot be more than 10^{-4} times as great. It represents an entirely new kind of astronomical object, whose existence was theorized by Hoyle, and a dozen or so are now known. There is much speculation as to their nature and origin, but to physicists their main interest is that deep inside them the gravitational field may be in Einstein's terms really strong and that, remote though they are, they may some day give us some hint of how such a field behaves.

CHAPTER 5

The Mössbauer Effect

There are several reasons in addition to that mentioned in the last chapter why it is desirable to have a precise measurement of the frequency of a beam of light, but any such measurement must be made by means of an instrument, since the human eye is completely unsuited for the purpose. Almost anyone can tell within an accuracy of 2 or 3 per cent which of two musical notes has the higher frequency, but any discrimination of frequencies by the eye is of the nature of a learned trick and cannot be nearly so exact. A variety of instruments are available to make up the defect, and they are generally of two kinds. The first is a spectroscope, an optical device that separates beams of light of different colors and presents them against a scale from which their frequencies or wavelengths (the two are closely connected) may be read. The precision is limited by that of the scale, and an accuracy of one part in 10^4 is very good indeed, although not nearly good enough for the experiment on gravity, where we are looking for an accuracy of one part in 10^{15}. However, it would be pointless to aim at an instrument capable of giving a reading to fifteen figures when what we need is much more modest, a device to measure the difference between two frequencies that are known to be very nearly the same.

The measurement can be made in the same way that one tunes a radio in order to cut out thousands of unwanted programs and reproduce only one—by the use of the principle of resonance. Resonance is a phenomenon that is often encountered in sound. If, for example, one holds down the damper pedal of a piano and shouts "oo" at the strings, the piano will respond with "oo" in a voice recognizably like one's own, because some of the freely vibrating strings of the piano resonate to particular musical tones in the voice whose frequencies are the same as those of the strings. This resonance phenomenon has an analogue in light. If light of a certain frequency strikes an atom, the atom will resonate to it, which means that it will absorb light at that frequency. This absorption is highly selective in the sense that if the incident light is not of exactly the right frequency, the absorption will not take place.

Let us investigate this process in more detail. Suppose that a number of atoms are free in space, so that they are not interfered with by their neighbors. Each atom has available to it a definite set of possible energies, as shown in Figure 2–4. If an atom is in one of the higher energy states, it will generally make an immediate transition to a lower one, and thus it will almost always be found in its lowest energy state, called the ground state, unless something has happened recently to energize it. Since energy is conserved, the atom cannot simply jump into a state of lower energy with no other effect, because this would require energy to disappear. Energy must be emitted in the process, and it ordinarily comes out as a single photon of light. Thus as an energized atom jumps down from a high level to a lower one and perhaps to even lower ones, one photon at a time comes off until the atom ends up in its ground state.

How does an atom get into an excited state in the first place? It can get there by several means, among them a precise reversal of the process by which it comes down. If a photon of exactly the right energy strikes the atom in its ground state, it will be absorbed, and the atom will go up into one of the excited states. If the energy of the photon is exactly equal to

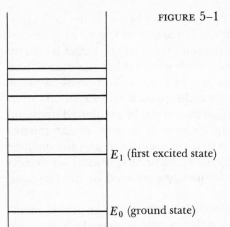

FIGURE 5–1 An atom in the first excited state undergoes a spontaneous transition to the ground state, emitting a photon that carries off the energy difference, $E_1 - E_0$. If a photon of exactly this energy strikes an atom in the ground state, it causes a transition to the first excited state.

E_1 (first excited state)

E_0 (ground state)

E_1 (Figure 5–1), the atom will go up into the *first* excited state. If it is nearly but not exactly E_1, the excitation cannot take place and still conserve energy, and so it does not take place at all. Since the frequency of a photon is proportional to its energy, comparing changes in energy levels is an accurate way of discriminating between frequencies.

The word "atom" is used rather loosely in describing these radiative processes. In fact, it is not only an atom that possesses a discrete series of possible energy states between which transitions occur with the absorption and emission of radiation. A molecule, which is a stable entity composed of several atoms bound together by forces of mutual attraction, has the same property, and so has an atomic nucleus, which is one hundred-thousandth the size of a typical atom. The only difference is that molecular energy levels are more closely spaced than those of atoms, whereas those of a nucleus are perhaps a million times as far apart. Thus the photons exchanged in these processes have energies very different from those of ordinary light, and their wavelengths are correspondingly different. Molecular radiations, if their wavelengths are so long that the eye does not respond to them, are called infrared rays, and nuclear radiations are called gamma rays.

The phenomenon of resonance can be put to use in the laboratory as follows: If, in the apparatus of Figure 5–2, the frequency with which the photon arrives at the target is exactly the same as the frequency with which it left the source, and if the target is identical in its chemical and physical properties with the source, then the photon will excite an atom (or molecule, or nucleus) in the target, be swallowed up there, and not go through to the counter. If for some reason the frequency is a little off, there is no resonance, and the counter will count. Thus the resonance technique is ideal for detecting the minute shift of frequency predicted by the principle of equivalence.

But there is a complicating factor. Light carries momentum as well as energy, and the atom that gives the light its bit of momentum suffers a recoil in doing so exactly like the recoil of a gun that fires off a bullet. The recoiling atom contains a certain amount of energy of motion that comes, of course, from the light itself, which caused the recoil. The light therefore has lost a small amount of its energy and starts off with less than it would have had if the recoil had not taken place. It arrives at the target with slightly too low an energy to permit it to excite any of the atoms there, and so it goes straight through. Even if it arrived with the right energy, it

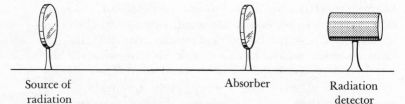

Source of radiation

Absorber

Radiation detector

FIGURE 5–2 Atoms of an emitting substance are excited by the addition of energy (in some unspecified way). The emitted quanta may pass through an absorber and be detected. However, if resonance occurs, the radiation reaching the detector is sharply decreased.

still would not excite an atom, because in entering and being absorbed it would communicate its momentum to the target atom. The recoiling target atom would therefore subtract some energy from the photon, and not enough would be left to excite it. This double loss is enough to make the experiment fail. How, then, can it be compensated? Clearly, it is quite hopeless to perform the experiment with atoms in a gas, because they are free to recoil. On the other hand, although particles embedded in a solid mass are not so free to recoil, they interact with each other so strongly that the energy levels are very much blurred; each atom does not have any one energy in particular, being continually pushed one way and another by the atoms on each side of it.

The fact is, though, that the nucleus of an atom is in many ways essentially a free particle. It is so well protected by layers of electrons that it scarcely senses the neighboring atoms at all. The nuclei in a solid, therefore, have perfectly distinct energy levels, just as in Figure 2–4, and we can use nuclear radiations that are the exact analogues of the atomic radiations. The trouble is that the nucleus itself will ordinarily waste some energy too, even though it is not perfectly free to recoil. Suppose that the tendency to recoil is not great enough to break a single nucleus loose and send it back through the rest of the sample like a bullet. Instead, the nucleus starts to vibrate in its place as a result of the recoil. How then does it lose energy?

Let us think about what happens when one taps on a solid piece of wood with a hammer. The hammer's energy of motion disappears suddenly. Some of it goes into sound in the air, but most of it goes into a wave of vibration, that is to say, a sound wave, which travels around inside the wood and is ultimately dissipated as heat. This shows that we have to consider the emission of sound waves from a recoiling atom in the solid.

We noted earlier that all wave fields have a particle aspect, although in many cases this aspect is not discernible by ordinary means. The particle aspect of sound, for example, is usually very inconspicuous. A particle of sound is called a *phonon,* by analogy with the word photon, which denotes a particle of light, and an audible sound consists of so many millions of phonons that its discrete nature is entirely imperceptible. But we are concerned here with the recoil of a single atom in a solid, which is a very much smaller circumstance than a tap with a hammer and which radiates very many fewer phonons. The essential fact is that phonons, like any other field quanta, come in integral numbers—the nucleus will radiate zero, one, two, three, etc., phonons into the solid and lose its energy that way. (Occasionally, too, the recoil results in the absorption of a phonon from the solid, but this process will not interest us here.)

The number of phonons emitted follows a well-known statistical law—first written down by the French mathematician Siméon Poisson—which governs rare and uncorrelated occurrences. To illustrate this law, we might use the number of personal letters someone receives per day. Suppose that none of his correspondents writes very often. Say that he receives, on the average, three letters a day; there are some days when he gets three, but there are other days when he gets one or two or four, and there is an occasional day (about one in twenty) when he gets none at all. This last does not mean that he is being neglected by his friends but only that a statistical event has happened. There will come a day when the postman will arrive with letters to make up for it.

What is important is the possibility of radiative processes in which no phonons are emitted. The most probable numbers of phonons emitted by a nucleus in a solid are zero and one; the emission of two or more is rare. If the atom recoils with the emission of a single phonon, it loses energy by this act, and the frequency of the accompanying photon is changed so much that it does not affect a target nucleus in the least. Therefore, the only quanta of radiation not counted in the experiment come from atoms that happen to have radiated no phonons. These atoms effectively do not recoil at all; that is, they do not lose any energy at all by recoil. With the zero-phonon processes we can eliminate the complications due to recoil and use the resonance effect to detect very small changes in frequency.

This possibility, although it involves scientific principles that have been well known for more than forty years, was not thought out because it never occurred to anybody. It was discovered experimentally and by chance by a young German physicist named Rudolf Mössbauer at Heidelberg University in 1958. He was observing resonance in nuclear radiation and noticed that the amount of radiation absorbed by the target was larger than it should be. One difference between a good scientist and a bad one is that a good scientist has a sense of the unexpected, and when he finds something happening in a peculiar way, he tries to see whether there is anything significant about it.

Mössbauer's experiment consisted in changing the temperature of the source and the target. The cooler they are, the fewer phonons are present, and the fewer are produced. Therefore, resonance occurs more often, and the target becomes less transparent to the photons. For his discovery, so rich in potential, and for its correct explanation, he was awarded the Nobel Prize for Physics in 1961.

Almost at once a number of people realized that they could now distinguish between frequencies to an accuracy of one part in 10^{15}, as necessary for the gravitational experiment. This experiment was performed throughout most of 1960 by Professor Robert V. Pound and his student, Glen Rebka, at Harvard University, where they used a tower 76 feet high. They were able to detect the shift in frequency predicted by the principle of equivalence and to show that it agreed with the predicted value to within 1 per cent or so.

At the end of 1961 a new application of the resonance principle to verify the principle of equivalence was announced. This time the light path was not from the bottom to the top of a 76-foot tower but from the sun's surface to the earth. The anticipated effect had heretofore eluded detection because it was masked by others much larger. In the first place, the radiant atoms at the sun's surface are continually undergoing collisions, which blur and distort their energy levels, and in addition the gases at the sun's surface are in turbulent motion at great speeds, which introduces substantial Doppler changes of frequency. On the other hand, the sun's gravitational field is so intense and the distance so great that the expected effect is much larger than predicted, being two parts in 10^6 instead of one in 10^{15}. Working at the Observatoire de Meudon, J. E. Blamont and F. Roddier produced resonance between light from strontium atoms in the sun's atmosphere and identical atoms in the laboratory. The anticipated shift toward lower frequencies and longer wavelengths was found, and it agreed with the theoretical estimate within the limits of accuracy of the observations, about 10 per cent. Neither of these experiments is accurate enough to be regarded as critical evidence for the principle of equivalence, but both represent a new method of attack on the problem of gravitation.

Actually two things have been accomplished—in the first place an experimental technique of unparalleled sensitivity has been discovered, and in the second it has been used to check one of the predictions of the theory of relativity, which had entered a long sleep in 1915 simply because most of the effects it predicts are so small. These and several other less spectacular experiments lead one to hope that we are entering a new and interesting phase in physics, which will be made possible by other ingenious discoveries, extending our fields of measurement many, many orders of magnitude beyond any ever seriously considered before. We have here a single, very restricted scientific phenomenon called the Mössbauer effect. One can do no more with it than one's ingenuity suggests. After all, what can be done with a measuring tape? It is good for nothing except the kind of measurement in which one puts one end here and the other end there and reads off a number. Yet it can be applied to creative purposes. Without a measuring tape the cathedral of Chartres could not have been built.

In 1958 Giovanni Cocconi and Edwin Salpeter of Cornell suggested that it was time to look for the anisotropy of inertia, small though it must be. Their idea was based on the fact that, as we shall see in Chapter 8, a nucleus can be made to vibrate so that it alternately lengthens and shortens in a certain direction. The frequency of any vibration of this kind depends on the mass in motion, and if the mass depends on the direction of the motion, the frequency should be affected. Cocconi and Salpeter presented arguments of a qualitative nature indicating that probably one part in 10^9 was the upper limit for the anticipated change in mass. Experiments performed at the University of Illinois showed that the anisotropy in mass, if it exists, is less than one part in 10^{15}, that is, smaller by 10^7 times than Cocconi and Salpeter's estimated upper limit. At about the same time, a group at Yale, who had read Cocconi and Salpeter's paper, realized that a different nuclear technique, called nuclear magnetic resonance, could be used to make the particular measurement even more exactly than the Mössbauer effect. They found that the effect must be less than one part in 10^{20}. This result is disappointing. We dislike concluding that there is no effect whatever because life would be so much more interesting if there were one. Hence we remain hopeful and say that it is perhaps just barely too small to observe. But to have brought the matter to this degree of accuracy within a few years after the publication of the first paper on the subject is a most remarkable achievement.

Although mathematics stands in the way of our knowing many of the consequences of Einstein's general theory, the equations seem to predict that inertia will be isotropic no matter how the mass of the universe is distributed. Thus the most important result of these experiments may be a check on the theory itself.

One other type of experiment can be done with the Mössbauer effect. We have suddenly discovered that the nucleus is a sensitive little instrument buried in the heart of an atom. If we can persuade the nucleus to respond to what is going on in there, we have a delicate way of finding out what the place is like. One recent experiment using a radioactive species of iron measured the strength of the magnetic field in the middle of an atom when the iron was magnetized. It turned out to be exceedingly large, 333,000 gauss (about ten times the field produced by a large electromagnet), and it is directed oppositely to the magnetization of the iron as a whole. There is at present no explanation of this value, a fact that makes the experimental discovery the more valuable.

The Mössbauer effect illustrates what may be an important part of physics in the second half of the twentieth century. It consists in being able to make precise measurements of things that people had always thought impossible to measure. For example, in recent years two groups have gone back and tried to measure the charge on a hydrogen molecule. We have been taught that a molecule has no charge because it contains equal numbers of positively and negatively charged particles. But what happens if the positive particle does not have exactly the same size of charge as the negative particle? In 1959 H. Bondi and R. A. Lyttleton, two British astronomers, pointed out that if there were an excess of about one part in 10^{18} either way, all atoms would tend to push apart and that this would explain the expanding universe. It is now known that the charge on an electron and the charge on a proton are the same to within one part in 10^{19}. This does not mean that there is no difference at all, but only that the difference, if it exists, is very small, and unfortunately too small to save the British theory.

The other extreme of contemporary physics is very largely qualitative, consisting in an effort to comprehend, even in a very rough way, some facts of nature that are entirely new to us at present. The most typical examples are to be found in nuclear physics, and particularly in elementary particle physics, where errors of 50 per cent are tolerated if they cannot be improved upon. The next chapter will deal with some of the results obtained in high-energy nuclear physics and the ways they affect our understanding of what goes on at the most fundamental level of nature.

CHAPTER 6

Elementary Particles

Of the different branches of modern physics, the study of elementary particles is
being pursued with the most energy and
zest. It is perhaps encouraging that this
is one field of study for which no utility
in terms of practical life or military strat-
egy is envisaged, now or for the future,
even though history shows us that most
scientific discoveries find their way into
technology. If the elementary particles do
not serve any useful purpose in our society,
let us try to see what part they play in our
ideas, for then it will be much easier to
grasp the nature and results of this complicated endeavor.

We have encountered in previous chapters four kinds of particles that today are
called elementary; they are electrons, protons, neutrons, and photons. (Phonons are
not fundamental because they belong to
waves of vibration in a substance composed of other particles.) Electrons and
protons, it will be remembered, are the
basic materials of the hydrogen atom. If
one combines neutrons and protons in
suitable proportions, nuclei are formed;
adding electrons to these produces atoms
of more or less complicated structure.
Finally, photons are the particles associated with the electromagnetic field of
light. We and our surroundings are composed of electrons, protons, and neutrons,

which can be thought of as the permanent substratum of all natural phenomena.

The role of photons is something different, and let us consider for a minute how they fit into the general picture of elementary particles. The electromagnetic field plays two quite distinct parts in the ordering of nature. First, light is an electromagnetic field, and it is as the quanta of light that photons make their appearance in physics. On the other hand, the first properties of the electromagnetic field to be studied had nothing to do with light but were concerned rather with force, for electric fields exert force and so do magnetic fields, and forces that are primarily electric in character hold together the atoms (though not the nuclei) of which everything is made. Thus the photons in the electromagnetic field are, in a general way, connected with the forces that exist inside atoms. We can say that the main part played by photons in nature is to sew electrons and nuclei together into atoms, and atoms into molecules, and finally molecules into solid substances. The use of the word "sew" here is, of course, figurative, but one way to think of the electric field of force between two charged particles is to imagine that the particles are continually swapping photons back and forth. However, to take this picture too literally probably leads to more confusion than enlightenment.

There is a diagrammatic way, introduced in 1949 by Richard Feynman, now of the California Institute of Technology, of visualizing processes of this kind. Feynman diagrams are of great use in guiding the mind through the mazes of field theory, and they cover the blackboards of theoretical physicists.

Suppose an electron and a proton pass each other. Their brief interaction may consist in the exchange of a single photon, as in Figure 6–1. The photon line is labeled by the Greek letter gamma (γ), and it is not necessary to specify its direction. Other possibilities, involving two photons, are shown in Figure 6–2. And there is nothing to keep events like those in Figure 6–3 from happening as well.

It becomes clear why field theory is such a complicated

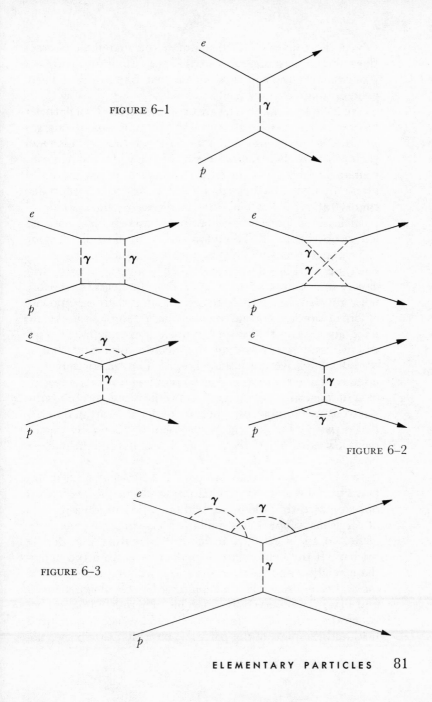

FIGURE 6–1

FIGURE 6–2

FIGURE 6–3

subject. Fortunately, in calculating electromagnetic interactions, the simplest diagrams are the most important, and even if one ignores every process but the first, one can get a fairly good estimate of the forces.

We have seen that nuclei are composed entirely of particles having either positive charges—the protons—or no charges at all—the neutrons—and that since all the protons repel each other by their mutual electric force while the neutrons remain unaffected, we have still to understand why the nuclei do not fly apart. One possibility is that, at the very short distances (about 10^{-13} cm) that characterize the spacing of particles in a nucleus, the ordinary laws of electricity no longer operate at all. This guess can be checked directly, for one can perform collision experiments in which charged particles are fired in such a way that they must pass extremely closely to one another; it is found that the guess is probably not right. It is much more reasonable that there is some sort of force holding a nucleus together that resembles the electric force attempting to disrupt the nucleus except that it is attractive and stronger than the electric force and that it operates only over very short ranges. The electric force exerted by a nucleus reaches out thousands of times the nucleus' own diameter into the space around the atom and binds the electrons to the nucleus, whereas the force of attraction between particles within the nucleus can be shown by experiment to vanish entirely or very nearly just beyond the nuclear boundary.

In 1935 it occurred to the Japanese physicist Hideki Yukawa that the uniformity of nature would be nicely served if one supposed that there is a field operating inside a nucleus at very short range that is similar, at least in some of its properties, to the electromagnetic field operating outside the nucleus. If the field could be shaken loose from the immediate neighborhood of the particles on which it acts, it would move through space like the quantized electromagnetic field and like it be observed as a particle. There is a relation between the range of the forces produced by such a quantized field and the mass of the particles: basically, the shorter the range, the greater the mass. The electromagnetic field has no effective limit on its range; the magnetic field of the earth extends thousands of miles into space, and accordingly we conclude that the mass of the photon is most likely zero. Yukawa's particle, on the other hand, had a mass that was estimated from the known range of nuclear forces to be of the order of 200 or 300 electron masses.

It is assumed that the electromagnetic field still operates in a nucleus to disrupt it but that the Yukawa field, within its limited range, exerts an attractive force that more than compensates the tendency to repulsion. For example, imagine that it is somehow possible to split a nucleus into two parts and separate them a little (Figure 6–4). The Yukawa force, which is strong only over short distances, now acts only very weakly to hold the two fragments together, while the electric force still pushes them apart. Therefore, they tend to separate further, faster and faster, finally rushing apart with great energy. This is the process known as nuclear fission, which produced the first atomic explosions and is now generating power in several countries. The energy released in fission is purely electric in origin, coming from the electric repulsion of the nuclear fragments.

FIGURE 6–4 Steps in nuclear fission. Only a few heavy nuclei can be split easily, but once the process begins, it is brought to completion by the mutual electric repulsion of the positively charged fragments.

Yukawa's paper was greeted with some reserve by physicists, although the idea that it presented was surely a very challenging one, and physics entered a difficult period, not to mention a war, while it was determining whether Yukawa's postulated particle actually existed. By the end of the war, it was clear that it did. It had been found in the debris

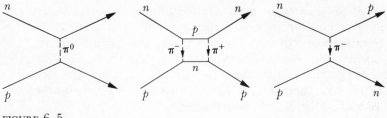

FIGURE 6-5

of atomic nuclei disrupted by cosmic rays streaking in from outer space, and later it was created in the laboratory. This particle has been given the name *pion,* and it is represented by the Greek letter pi (π). A pion's mass turns out to be about 273 times the mass of an electron, well within the range predicted by Yukawa. Another property of the pion, also correctly predicted by Yukawa, is that it is unstable, for it tends to disintegrate into other particles after an interval measured in hundred-millionths of a second; thus a pion cannot be considered as a permanent inhabitant of the universe in the sense that an electron, neutron, or proton is or a very nearly permanent inhabitant like a photon, which may have started from a far distant star ten billion years ago and now be flying toward the earth.

The Feynman diagrams for the interaction of a proton and a neutron (Figure 6–5) are similar to those describing the electromagnetic interaction and others more complicated. The last one illustrates a phenomenon called exchange, well known in experimental physics, in which the neutron becomes a proton and vice versa.

Figures are somewhat deceptive in a discussion of entities as tiny as these, but some indication of the strength of the forces that pions exert inside a nucleus is given by the fact that the electric repulsion on a single proton in a nucleus of uranium is around 50 pounds, and the pion field must be more than strong enough to counteract this. If it were not for the pion field, the universe would be filled with a clutter of wandering unstuck particles and a few hydrogen atoms and molecules. Thus electrons, protons, neutrons, photons, and pions are, so to speak, enough for nature as we perceive it, and a universe composed of these five types of particles would, as far as we know, have most of the properties of the universe we study. But there are more particles, and perhaps the most baffling puzzle of modern physics is to understand why there is this superabundance and how the new particles are related to the five just mentioned.

PROPERTIES

We have noted that it is possible to measure the mass, charge, and spin of an elementary particle; a further quantity must be added to this list—its lifetime. As we have seen, a pion free in space lasts but 10^{-8} second, after which it decays into smaller particles. A pion has no spin, but its mass and charge have been determined. The mass is about 273 times that of an electron, and positive, negative, and neutral pions have been found.

Positive and negative pions are apparently twins, identical except for their charges. Their masses are equal (to the greatest accuracy of which measurement is capable), their spins are both zero, and the lengths of time they last before disintegrating into other particles are the same. A neutral pion is at first sight more like a cousin—its mass is 264 electron masses, and its lifetime 10^8 times as short as that of a charged pion. What makes it appear that the three particles are fundamentally of the same kind is that all three seem to contribute interactions of equal strength in keeping a nucleus together.

It was mentioned in Chapter 1 that the total electric charge in the universe is not altered by anything we can do to it. This general principle of physics is known as the conservation of charge. Thus if a positive pion is produced in some nuclear explosion, we can be sure that a precisely compensating negative particle (it may be a pion, or it may be something else) is produced at the same time. Every type of elementary particle seems to have some sort of charge symmetry.

One of the puzzling features of Dirac's relativistic wave equation for electrons, discussed in Chapter 3, was that the equation has two sets of solutions, and when properly examined, half of these solutions correspond to particles of negative charge like the known electrons, and half of them to unknown particles of positive charge. Specifically, Dirac's equation gives only the value of the square of the electric charge of the particle it describes, and as one knows from elementary algebra, a given quantity has two different square roots, a positive one and a negative one. Dirac's first thought was to throw away half the solutions of his equation, those not corresponding to electrons. But there are a number of mathematical operations requiring all the solutions of an equation, and if half of them were missing, one would not know how to proceed. For a while an effort was made to identify the positive particles with protons, which were also known at the time, but it was clear from the structures of the solutions that the positive particle had to have the same mass as the electron, or else the theory was false. It is gratifying to learn that the unanticipated and undesired solutions of Dirac's equation do in fact correspond to something physically real, just as the desired solutions do.

Five years after the publication of Dirac's theory, positive electrons, as they were then called, were discovered in the debris of atoms disrupted by cosmic rays. These particles have been given the name positrons, and it is also useful sometimes to refer to them as antielectrons to emphasize their relation to electrons. More recently, antiprotons have been produced in the laboratory that are identical with protons except for their charges.

The question of whether a neutral particle has an antiparticle associated with it is at the moment a difficult one, because if a particle and antiparticle differed only in having opposite charges, there would be no obvious means of distinguishing a neutron from an antineutron. Yet antineutrons are known to exist. One convenient way of characterizing them is in terms of their magnetic properties. We have noted that an electron acts not only like a little charge but also like a little magnet, and the same is true of all particles that have spin. Even the neutron, which has no electric charge, is found to have a small magnetic effect, similar to that of a tiny bar magnet oriented parallel to the axis on which the neutron is spinning. This property distinguishes the neutron from the antineutron, for with an antineutron the magnet is oriented in the opposite direction. The neutral pion, on the other hand, has no spin, and consequently this method for distinguishing particle and antiparticle does not apply for it. Apparently a neutral pion is identical with its antiparticle.

We look for symmetries in nature because symmetries are properties of simplicity and give us an agreeable feeling that all is well. It was unexpected but pleasing to find a symmetry between one charge and its opposite, and so if atoms made up of electrons, protons, and neutrons have certain properties, then atoms made up of positrons, antiprotons, and antineutrons would be expected to be identical to them in all respects except for the charges of their particles. Now the universe as we encounter it in our investigations does not show on a large scale this symmetry that we observe on a small scale, for all the atoms of which you and I are composed are of one kind; the other kind has never been discovered. Thus nature seems to have open to it a type of symmetrical behavior that it has not used, and this sets us a puzzle that must someday be solved.

Part of the answer is quite obvious from another property of elementary particles. It is possible to produce positrons in a cosmic ray explosion, and indeed a typical one will yield a number of pairs of positrons and electrons, thus satisfying the law of the conservation of charge. The inverse reaction is also possible. If in the course of time a positron and an electron come close together in space, they will annihilate each other with the emission of a flash of light and totally disappear; the same is true for protons and antiprotons and for neutrons and antineutrons except that their annihilation results in pions.

Let us now imagine some cosmic process of creation in which great numbers of protons and antiprotons, neutrons and antineutrons, and electrons and positrons are produced and begin to wander around. Ultimately they will sort themselves out into atoms and antiatoms, with few unattached particles left over, and all this time collisions will be taking place. When an atom and an antiatom collide, the electrons and positrons will annihilate each other, and so will the nuclear particles, and the atoms will disappear in flashes of light and pions. Similarly, when free particles find their antiparticles, they will also disappear in pairs. If the original process of creation were somewhat one-sided, this process of mutual annihilation would continue until all of one set of particles had disappeared. The remaining set would then be entirely stable from this standpoint, for there would be nothing left that could annihilate it.

We are at liberty to suppose that the process by which the universe began was asymmetrical in this way. Speculations about this remote event are perhaps not profitable from a scientific point of view, but we can also at least conjecture that an even-handed symmetry may have been shown in the beginning of things. We could explain the present asymmetry by assuming that although perhaps equal numbers of particles and antiparticles of each kind were originally produced and may still exist in the universe, their distribution may be somewhat irregular and that the preponderance of electrons, protons, and neutrons in the part of the universe with which we are familiar may be balanced by an equal preponderance of antiparticles elsewhere. It is possible to imagine a space voyage at the end of which one approaches a pleasant and inhabited planet giving every sign of a busy and happy life; one enters the atmosphere and finds, alas too late, that it is not a planet but an antiplanet and that its atmosphere is nibbling away at the spaceship and destroying it and its passengers through the process of annihilation. Travel to distant parts of space seems to be forever out of the question, but if the barriers to it, which now appear insuperable even in theory, should be removed, a space explorer of the future will carry with him a little pellet that he will fire toward the planet he wishes to explore in order to see whether it vanishes as it enters the planet's atmosphere.

In addition to the physical properties already mentioned that distinguish one elementary particle from another, a somewhat more sophisticated property is very important. If a large number of particles crowd together to form a single entity, as a nucleus and a number of electrons form an atom, then certain considerations come into play limiting the possible energies that the particles can have. With regard to the structure of atoms, for example, a rule states very roughly that no two electrons in an atom can have exactly the same energy. Since things in nature tend to seek the lowest energy available to them, if one imagines putting together an atom by pelting a nucleus with electrons, the first electrons arriving will fall to the states of lowest energy, and subsequent electrons, being unable to enter these states, will have to occupy states of higher energy, which situate them farther out from the nucleus; in this way an atom will be built upward and outward, and different electrons in it will turn out to contribute different properties to it. If the exclusion principle, as it is called, did not apply, all the electrons in an atom would be in the same lowest state, and the different kinds of atoms would be far less sharply distinguished in their chemical and physical properties than they are.

The question of why some elementary particles obey the exclusion principle is not entirely answered today, although the existence of the principle can be related to other and more fundamental facts. But one can ask whether every kind of particle obeys this principle. The answer is no. Roughly speaking, it is the heaviest and the lightest elementary particles that do and the particles of intermediate mass that do not. Particles without spin do not, and this property extends even to composite systems.

For example, there are two isotopes of the gas helium. In one, helium-4, the nucleus contains two neutrons and two protons, and the spins of these are directed in such a way that they compensate and the nucleus is without a spin. Also, the spins of the two electrons are oppositely directed so that they compensate, and the entire atom is therefore spinless. In the lighter isotope, helium-3, such a compensation is not possible because there are only three particles in the nucleus and there is no way of arranging the three spins so that their effects add up to zero. Thus, although the electron spins still compensate, there is a nuclear spin, and the whole atom of helium-3 is found to obey the exclusion principle. Although this difference in helium isotopes has no great effect on the other physical properties of isolated atoms of helium, we shall see in Chapter 8 that the bulk properties of helium when large numbers of atoms congregate together at low temperatures in a liquid are very drastically influenced by it and that liquid helium-3, although chemically identical with liquid helium-4, has very different physical properties.

One final remark should be made about other possible properties of elementary particles. If they were truly elementary, it might be reasonable to suppose that the list of properties we have given is exhaustive. But there is no special reason to assume that they are, and such features as the instability of some of them and their tendency to break into others are perhaps good reason for assuming that the situation is somewhat the same as it was when the word "atom," which in Greek means "that which cannot be cut," turned out to have been inadvertently applied to a complicated structure of electrons, protons, and neutrons.

Let us suppose for a moment that the particles which we call elementary are, in fact, in some sense composite—that they are structures of fields which are more elementary still. If this is true, then there is one further property that we might expect these particles to manifest. Both atoms and nuclei (which we have discussed), and molecules, which are clusters of atoms, have the property that, together with the stable state in which they are usually found, they have available to them a certain set of unstable, excited states into which they can be put by a suitable process involving the application of energy. A test of whether a particle is really elementary would then be to see whether there is any way of energizing it into some short-lived state of excitation from which it would, like a molecule, an atom, or a nucleus, shortly drop into a state of lower energy. This loss of energy, of course, would have to be compensated by the appearance of energy somewhere else. In atoms and molecules the energy is most likely to appear in the form of photons. Indeed, the photons that we see around us have precisely this origin in the sun, or the gas of a fluorescent light, or the hot wire of an incandescent bulb. Correspondingly, nuclear de-excitations are accompanied by the emission of particles from the nucleus, some of which are photons of such short wavelengths that they do not affect our eyes but pass right through us. The de-excitation of an elementary particle could thus be expected to result in the emission of something, possibly a photon and possibly not. One of the principal projects of the physics of elementary particles is to investigate the existence of the states of higher energy. At the time this is being written, a number have been found. Their ephemeral occurrence does not prove that any of the particles from which they are excited are composite in nature, but it makes it plausible. On the other hand, it may well be that at this scale of smallness it is really meaningless to try to distinguish between what is elementary and what is not. The point is, perhaps, not a very difficult one, and it may be clarified in a few years.

CREATION AND DESTRUCTION

It is a generally accepted principle of nature that matter can be neither created nor destroyed. If one looks out his window in the morning and sees that his car is no longer outside, he deduces only that someone must have taken it away while he slept. The possibility of the car's simply disappearing does not enter his mind. But we have just referred to events that apparently contradict this principle of nature—the creation and annihilation of electron-positron pairs and other forms of matter. Thus the principle of the conservation of matter seems to be violated on the microscopic level though not on the level of ordinary proportions.

It is possible to make a general statement of conservation that covers all known cases if we include energy and matter in the same law. Einstein showed that to matter corresponds energy and to energy corresponds matter and that these two currencies, superficially so different, may be exchanged at a rate given by

$$E = mc^2$$

where E is energy, m is mass, and c is the speed of light. The most important characteristic of this equation is that, since light travels very fast, c is a very large number and c^2 correspondingly much larger; therefore, the exchange rate is heavily on the side of mass: a small amount of mass equals a large amount of energy. It is difficult to suggest the magnitudes involved, but if the mass of a drop of water could be completely converted into energy at this rate in a gentle and non-explosive manner, the resultant energy would be enough to lift several 10-ton trucks onto the moon. Normally, exchanges of energy are not accompanied by observable changes of mass simply because the amounts of energy encountered in daily life or in the laboratory are not large enough to have perceptible amounts of mass associated with them. Still, a hot flatiron weighs, in principle, a bit more than a cold one, and any moving object gains extra mass from its energy of motion.

At the level of elementary-particle phenomena, the situation is different, in that at the energies at which these reactions normally occur the increases and decreases of mass are perfectly perceptible; in fact, they often play a dominating role. From this point of view let us follow the events that take place after a cosmic ray from outer space strikes the nucleus of an atom high in the atmosphere with terrific violence and produces a downward shower of particles. The energy of the incident particle, which might be great enough to throw a tennis ball over a small house if it could be directed to that purpose, reappears first of all in the tremendous energy of motion of the various particles that emerge from the disrupted nucleus. We find most or all of the original neutrons and protons in the nucleus speeding downward and separating from one another, but also in the debris we find pions and pairs of electrons and positrons and possibly antiprotons and antineutrons as well. Part of the energy liberated in the impact has gone into the creation of pairs of particles out of nothing but energy.

In a very short time, the pions produced in this explosion begin to decay. The decaying process consists of the disappearance of the pion and the appearance of two new particles, one called a muon, which is represented by the Greek letter mu (μ), and the other called a neutrino, which is represented by the Greek letter nu (ν). The mass of a muon at rest is about 207 times that of an electron, and its charge may be negative or positive. A neutrino cannot be brought to rest, but it would have no mass if it could. A pion at rest has a mass about 273 times that of an electron. Thus if it decayed at rest, the neutrino and the muon would have between them an extra mass equal to that of 66 electrons, which corresponds to their energy of motion. But a fast-moving pion in a cosmic-ray shower may have a mass several times 273, so that the mass-energy shared between the muon and the neutrino may be much more. After some 2×10^{-6} second, the muon in turn decays, into an electron, a neutrino, and an antineutrino. A neutrino is related to an electron in the sense that it has the same spin and obeys the exclusion principle; it differs in having no charge and, as far as is known, no rest-mass. Neutrinos and antineutrinos are distinguished by the directions of their spins at the moment of formation; when an antineutrino is formed, it speeds off spinning in the manner of a screwdriver pointed in the direction of its travel, whereas a neutrino spins in the opposite direction. In all other respects they are identical. The electron and the two neutrinos issuing from the decay of the muon are, as far as anyone knows, stable. The electron finally becomes part of an atom somewhere, while the two neutrinos rush off into space with a velocity equal to that of light, and since their interactions with matter are exceedingly weak, for millions of years they pass through anything in their paths and come out the other side.

The decay of a muon brings us to a new kind of Feynman diagram, Figure 6–6, in which we use a bar to indicate an antiparticle. The previous diagrams have consisted of vertices at which only three lines come together; here there are four. Neutrinos also issue from four-vertices. Thus the decay of a pion may take place via the process shown in Figure 6–7 and others more complex. Theorists predicted that the μ^+ on

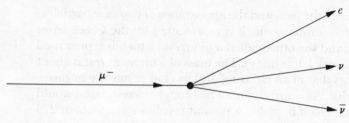

FIGURE 6–6

the diagram might sometimes be a positron, and a detailed calculation showed that this would be the case about once in 8100 decays. There was dancing in the cafeteria line at the European Center for Nuclear Research when the experimental confirmation of this result reached Geneva.

The neutrons emitted in the course of nuclear disruption also turn out to be unstable. We have learned that neutrons and protons are much alike except for their charges. More specifically, they differ slightly in mass; a neutron is some 1839 times as massive as an electron, whereas a proton is only about 1836 times as massive. Therefore, there is enough mass-energy available for a neutron to turn into a proton and an electron, and, in fact, a neutron becomes a proton, an electron, and a antineutrino (Figure 6–8) at the end of some 13 minutes of free life, a lifetime enormously longer than that involved in the decay of any other known unstable elementary particle. It is natural to ask whether a neutron inside a nucleus can take part in the same sort of decay. The answer is that in some atoms it can and in most it cannot. The reason why it cannot is rather complicated and depends upon the conservation of energy and the exclusion principle. When the decay is possible, the neutron inside a nucleus turns into a proton, while the electron and the neutrino are emitted

FIGURE 6–7

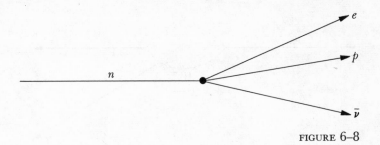

FIGURE 6–8

quite forcefully from the atom. This is the origin of one type of radioactivity, known as beta radiation, in which electrons leave a nucleus as it undergoes a change from one nuclear species to another. In most of the nuclei with which we normally deal, this decay process cannot take place, and for many years it was not even known that free neutrons break down as they do.

Theorists interested in preserving the uniformity of nature in the face of a rather diverse assortment of particles have proposed to explain the existence of Feynman diagrams with both three- and four-line vertices by inventing a new heavy particle, designated W, in terms of which muon decay, for example, would be given as in Figure 6–9. It is difficult to find the W experimentally, for very few would be produced in experiments at available energies, but there is some evidence that it exists—a theoretical argument and a large experiment. The experiment will be described in Chapter 7. The argument goes as follows. Consider the process diagramed in Figure 6–10, by which a muon decays into a neutrino, a positron (e^+), and a photon. It can be calculated that this decay should occur about once in 10,000 times.

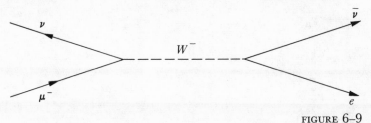

FIGURE 6–9

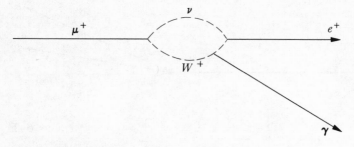

FIGURE 6–10

Careful studies, most recently at the Enrico Fermi Institute for Nuclear Studies at Chicago, have shown that it occurs, if at all, less than once in 20,000,000 times.

At this point, W seems to be in danger. But let us find a way to save it. One way is to suggest that there are two kinds of neutrinos, one associated with electrons and the other with muons, so that muon decay should be diagramed as in Figure 6–11. If the two are really distinct, then the decay $\mu^+ \to e^+ + \gamma$ is impossible, since the neutrino in the diagram would have to play both roles. The question now is to find the two neutrinos. This is easier than finding the W and was done in 1962 at Brookhaven, in an experiment to be outlined in the next chapter. The matter then stood as follows: to save the W, a far-fetched guess was made, which turned out to be correct. Is there a W? Experiments are not yet conclusive, and we still do not know.

The general experimental method applied to the study of unstable elementary particles is to create them and watch what they do. Most of the particles are extremely short-lived,

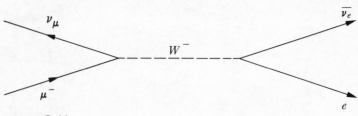

FIGURE 6–11

so that the processes of creation and of observation must be timed for almost the same instant. The particles are created by high-energy impacts and must be observed somewhere close to where the impact takes place. One possibility is to go to the top of the earth's atmosphere and wait for cosmic rays to strike. This is, however, inconvenient and inefficient, and the entire program is best carried out in a laboratory. Such a laboratory centers around a machine whose purpose is to give high energies to stable elementary particles. The particles are then smashed into some suitable atomic target, and finally observations are made of the various physical changes that occur at the moment of impact. The next chapter will tell how some of these experiments are performed and what the results are.

CHAPTER 7

Physics at High Energies

There are a number of laboratories in the world that exist largely in order to study the elementary particles. Notable among them are the Brookhaven National Laboratory on Long Island, run jointly by the major Eastern universities; the Lawrence Radiation Laboratory of the University of California at Berkeley; the European Center for Nuclear Research (CERN) at Geneva, run by a dozen European countries; and the Joint Institute for Nuclear Research at Dubna, near Moscow. The activities of each of them center around one or more of the machines known as high-energy accelerators. These are used to speed up stable particles of some sort, usually protons or electrons, to an energy so high that when they are finally made to collide with a target, the energy released in the impact will produce new elementary particles for study. Associated with the accelerator is a building full of apparatus for monitoring and controlling its function and for observing and measuring the events taking place at the target. The typical high-energy accelerator (Fig. 7–1) involves a circular racetrack—an evacuated ring-shaped tube—in which charged particles whirl round and round while they gradually gather energy. The particles are guided by a magnetic field, which acts on a charged particle in such a

FIGURE 7–1   Cosmotron, a synchrotron that accelerates protons in a circular path to speeds
approaching that of light. (From Brookhaven National Laboratory, Upton,
N.Y.)

way as neither to speed it up nor to slow it down but merely
to bend its path into a circle. Thus the greater part of a high-
energy accelerator consists of a large magnet; the one in the
Alternating Gradient Synchrotron at Brookhaven is in the
form of a circle some third of a mile in circumference. Inside
this magnet is the evacuated tube, and at one place in the
circuit there is a region where an electric force can be applied
so as to add a little energy to the particles each time they
come around, gradually building up their energy to a very
large value.

To give some idea of the energies required and available
in elementary-particle physics, we shall express them in the
unit of millions of electron volts, abbreviated Mev. It takes
about 1 Mev of energy to create an electron-positron pair and

about 10^{13} Mev to make up a foot-pound (the work necessary to raise a 1-pound weight 1 foot). Since protons are some
2000 times as massive as electrons, ideally it should take
about 2000 Mev, or 2 billion electron volts (Bev* or Gev),
to create a proton and an antiproton. But actually 2 Gev will
not do it, because about twice this much energy is necessarily
consumed in the process of creation and in the energy of motion of the particles after they are created. Thus about 6 Gev
is needed before protons and antiprotons appear. The most
powerful accelerators now in operation are two synchrotrons,
at Brookhaven and Geneva, which produce protons of 33 and
28 Gev, respectively. At Serpukhov in Russia, a 60-Gev machine is under construction, which, when completed, will be
the largest in the world.

With a few Gev it becomes possible to create heavy particles entirely out of energy. However, the tentative picture of
elementary particles outlined in the preceding chapter suggests that there is another, easier way of creating things to
study, for atoms in excited states do not have to be created out
of energy; it is only necessary to start with unexcited particles
and add energy to them by means of a collision. Whether or
not the excited particles will behave like the original unexcited ones is open to question, but if a particle is in any sense
a composite system, it is at least reasonable to look for such
states. Thus we should expect two types of reactions to occur
at the target of a high-energy accelerator. One is the creation of particles purely out of energy; the other is the formation of excited states of existing elementary particles using
the energy released at the impact.

At the target of an accelerator, where elementary particles
are formed either by creation or by excitation, it is essential to
install detectors of various sorts so that one can observe and
measure what happens. To examine elementary particles one
at a time is a challenging problem. As was mentioned in

---

* Because Europeans and Americans use the word "billion" to denote different numbers, this ambiguous term is on the way out. It is being replaced
by *gigavolt,* as in "gigantic," abbreviated Gev.

Chapter 1, until the last sixty years of atomic physics there was no way of establishing contact with the world of atomic smallness even firmly enough to judge whether atoms existed at all. Einstein discovered that there is a bridge between the atomic world and the world of our senses in particles of intermediate size, which can be affected by the one and observed by the other. Since the initial discovery, new methods of studying elementary particles have been invented every few years.

The typical detector of elementary particles consists of a relatively large-scale physical system that is in some way unstable, just on the brink of undergoing a sharp change of state. Its condition is made so precarious that the arrival of a single atomic particle can tip it over and cause a change that is easily detectable by means of instruments. Many kinds of instability have been exploited for this purpose. Two basic types of instruments have resulted, one for counting particles and the other for watching what they do. The most familiar example of the first is the Geiger counter, which is an electrical discharge tube with such a high voltage applied to it that a single particle can start an electric current flowing in it. This device was invented in 1907, and there are by now many developments and modifications of the idea all of which, however, work in roughly the same way, giving their readings in terms of clicks or pulses of electricity that can be timed and counted. The pulse from the counter enables us to estimate the energy of a particle and to know the precise moment at which it passed through. Counters are therefore useful for measuring the extremely short intervals between the creation and decay of elementary particles.

There are other devices for observing elementary particles that give us a photograph of an entire process. The prototype of such instruments is the Wilson cloud chamber, invented in 1912. It works as follows: If moist air is cooled, a cloud forms. By a rather complicated chain of events, the excess moisture in the air suddenly begins to form tiny but visible droplets of water. The presence of electric charges encourages the formation of these droplets. Dust particles, for example, are charged, and hence dusty air forms clouds more easily than pure air; in fact, in very pure air one often has a situation known as supercooling, in which no cloud forms at all, even if the air is moist and very cold. When an electrically charged particle passes rapidly through air, it leaves behind a little train of slightly disrupted atoms which are electrically charged, and these, it turns out, are ideal for initiating the condensation of a supercooled vapor. If such a supercooled vapor is produced for a moment in the laboratory and a particle speeds through, at the next instant a little track of cloud appears, very like the track behind a plane at high altitudes and formed by much the same process. A moment later this track disperses, but a photograph taken at the right time enables it to be very sharply distinguished and measured.

Figure 7–2 is a cloud-chamber photograph of the decay of a pion into a muon and electron followed by the decay of the muon. Since the two neutrinos that accompany the electron in the decay of the muon are electrically neutral, they do not disrupt atoms as they pass through the chamber and therefore produce no tracks. This failure to picture every particle is an unfortunate feature of all cloud chambers, and it adds to the difficulty of physics at high energy.

There are two other disadvantages of a cloud chamber. First, it provides no way of measuring time intervals, and, second, if the process one wishes to study entails the collision of one particle with another, a chamber of moist air is not a good place to study it because air is principally empty space and often the kinds of collisions one wants are very unlikely to occur at all.

An enormous improvement over the cloud chamber is the bubble chamber, developed by Donald Glaser at the University of Michigan during the 1950's. The bubble chamber uses another instability, this time involving the boiling of a liquid. We are accustomed to the idea that boiling begins as soon as the right temperature has been reached. Liquids of immaculate cleanliness, however, can be heated somewhat above their boiling temperatures for a moment or so without beginning to boil, because the bubbles of vapor form

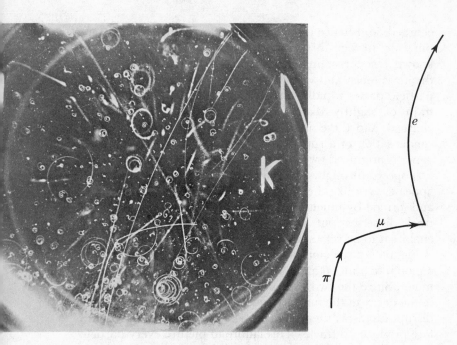

FIGURE 7–2 Cloud-chamber photograph of pion-muon-electron decay. (From W. Gentner, H. M. Leibnitz, and W. Boethe, *Atlas of Typical Expansion Chamber Photographs,* Pergamon Press, London, 1954.)

much more easily if there is some speck of impurity to encourage them. Disrupted atoms in the train of a charged particle act as impurities, and a photograph taken an instant after the passage of the particle shows a little track of bubbles. One advantage of the bubble chamber is that since liquids are far more dense than gases, the collisions in which a physicist is interested are much more likely to occur in it than in a cloud chamber. Another advantage is that pions and muons often travel so fast in a cloud chamber that they leave the apparatus without our being able to see what becomes of them; when such particles pass through a liquid, however, they soon slow down, and they generally come to a stop and decay within the chamber. Figure 7–3 shows the π-μ-e decay process in a bubble chamber.

Both cloud chambers and bubble chambers have the inherent disadvantage that they are sensitive during only a small fraction of the time. The rest is consumed by what is known as recycling—getting the instrument ready to fire again. Also, the bubble chamber cannot be triggered by an external set of Geiger counters arranged to select events worth looking at. This is because the liquid does not start boiling quickly enough to catch the particles that the counters detect, and so a bubble chamber must be set to take pictures automatically by the hundred or even the thousand, and all of them must be scanned afterward to see whether they contain anything interesting.

An important innovation that gets around this disadvantage (while introducing disadvantages of its own) is the spark

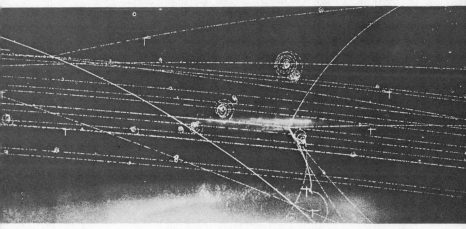

FIGURE 7–3 Bubble-chamber photograph of pion-muon-electron decay. The collision of two protons results in six new tracks. The four lower ones are made by pions, two of which continue to decay. (From Brookhaven National Laboratory, Upton, N.Y.)

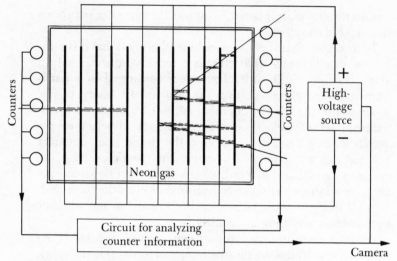

FIGURE 7–4 Spark chamber. A fast particle enters from the left, triggering a counter. A billionth of a second later, several counters on the right also fire. If this is the type of event desired, the analyzing circuit sends a pulse of voltage to a high-voltage supply, which briefly applies about 1000 volts between adjacent plates. Sparks flash from plate to plate in the path of the charged particle, and a camera photographs them.

chamber (Figure 7–4), of which the first workable model was made in 1959 by S. Fukui and S. Miyamoto of the University of Osaka. An airtight chamber contains a gas at low pressure and a series of very thin metal foils to which a high voltage can be applied after a particle has passed through and left a train of charged gas atoms behind it. Sparks jump from foil to foil along this track and can be photographed by their own light. Figure 7–5 is a spark-chamber photograph of a neutrino collision yielding a muon. This instrument was an essential part of the experiments that discovered the second neutrino and searched for the W particle. Recently a group in Russia, headed by A. I. Alikhanian, has been able to produce spark tracks almost 2 feet long, which are very sharp and clear in photographs and seem to follow accurately the paths of the particles that created them. The spark chamber may turn out to be a useful substitute for a cloud chamber.

It is, of course, instructive to have photographs of these hidden processes of nature, but it is the business of a physicist to measure, and the main technical problems connected with these instruments concern the numerical measurements that can be made with them. The interpretation of cloud-chamber and bubble-chamber photographs is a long job that is at present very largely mechanized but still involves much hard human labor. Nevertheless, when this labor is done, one finds it possible to infer the mass of a particle, its charge, and its energy at any point from examining the picture of its track.

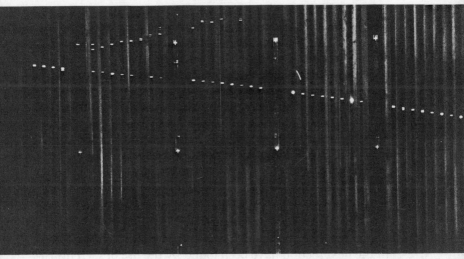

FIGURE 7–5 Spark-chamber photograph. The long straight track is that of a muon, produced when a neutrino hits a proton. The other track entering from the left is thought to be that of a gamma ray. (From Brookhaven National Laboratory, Upton, N.Y.)

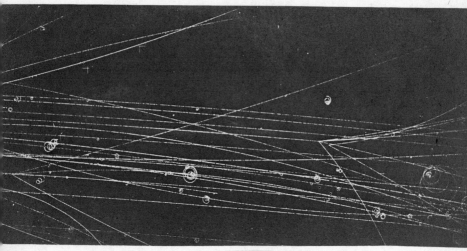

FIGURE 7–6   Bubble-chamber photograph showing
the tracks of pions produced when a
proton beam strikes an aluminum target entering from the left. Small circular
tracks appear where electrons, relatively
low in energy, have been knocked loose
by the high-energy pions. At the right
center, where several tracks leave the
primary path, a pion has collided with
a proton to produce some secondary
particles. (From Brookhaven National
Laboratory, Upton, N.Y.)

In this way particles can be identified, and anything new that
does not correspond to a known particle in any of these par-
ticulars can be spotted. Most of the particles to be discussed
in the next section have been found by the use of such instruments.

ELEMENTARY PARTICLES

The particles we have discussed so far have mostly been
known for many years, but more are being discovered all
the time. The known particles and their excited states

TABLE 7-1. ELEMENTARY PARTICLES

FAMILY	MEMBERS	BOSON OR FERMION	LIFETIME, SECONDS	DECAY
baryon				
intermediate boson*	W	B	10^{-16}	intermediate
hyperons	Ξ, Σ, Λ	F	10^{-10}	weak
hyperon resonances†		F	10^{-23}	strong
nucleons	n	F	10^{3}	weak
	p	F	stable	
meson	K, π	B	10^{-8}	weak
meson resonances†		B	10^{-23}	strong
lepton	μ	F	10^{-6}	weak
	e, ν_{μ}, ν_{e}	F	stable	
photon	γ	B	stable	

*If it exists.
†There are a great variety of resonances.

can be grouped in families as in Table 7–1, where the listing is roughly in the order of decreasing mass. The words *baryon, meson,* and *lepton* derive from Greek words meaning heavy, intermediate, and light. A *fermion* is a particle that obeys the exclusion principle, whereas a *boson* is one that does not. The general name for the excited states of elementary particles is *resonances.* We have encountered this term before. Used here, it reflects the fact that the particles are formed in collisions over a fairly narrow range of energies and avoids the question of whether they should be called "elementary." Tables 7–2 and 7–3 list the baryons and mesons and a few of their resonances,* and Table 7–4 lists the leptons.

The excited states of a nucleon (N) are given in Table 7–5.

*A table pretending to be complete would be long and would quickly go out of date. More data, together with a detailed discussion of some contemporary points of view, appear in G. Chew, M. Gell-Mann, and A. H. Rosenfeld, *Sci. American,* **210,** 74 (February, 1964).

TABLE 7-2. BARYONS

PARTICLE AND CHARGE	NAME	REST MASS, PROTON MASSES*	TYPICAL MODE OF CREATION†	LIFETIME, SECONDS	DECAY	DECAY PRODUCTS†
$\Xi^*(0)$	xi-star	1.63	$K^- + p \to K + \Xi^*$	$\sim 10^{-23}$	strong	$\pi + \Xi$
$Y_1^*(-,0,+)$		1.47	$K^- + p \to \pi^0 + Y_1^*$	$\sim 10^{-23}$	strong	$\pi + \Lambda^0$
Ξ^-	xi-minus	1.408	$\pi^- + p \to K^0 + K^+ + \Xi^-$	2×10^{-10}	weak	$\pi^- + \Lambda^0$
Ξ^0	xi-zero	1.402	$\pi^- + p \to 2K^0 + \Xi^0$	3×10^{-10}	weak	$\pi^0 + \Lambda^0$
Σ^-	sigma-minus	1.275	$\pi^- + p \to K^+ + \Sigma^-$	2×10^{-10}	weak	$\pi^- + n$
Σ^0	sigma-zero	1.270	$\pi^- + p \to K^0 + \Sigma^0$	$\sim 10^{-20}$	electro-magnetic	$\Lambda^0 + \gamma$
Σ^+	sigma-plus	1.268	$\pi^+ + p \to K^+ + \Sigma^+$	8×10^{-11}	weak	$\pi^0 + p, \pi^+ + n$
Λ^0	lambda	1.189	$\pi^- + p \to K^0 + \Lambda^0$	2×10^{-10}	weak	$\pi^- + p, \pi^0 + n$
$n(0)$	neutron	1.00138	nuclear disruption	1010	weak	$p + e + \bar{\nu}_e$
$p(+)$	proton	1.000	$n \to p + e + \bar{\nu}_e$		stable	

* Proton mass = 1836 electron masses.

† Where charges are not specified, any charges that add up to the right amount are possible.

TABLE 7-3. MESONS AND THEIR RESONANCES

PARTICLE AND CHARGE	NAME	ANTIPARTICLE	REST MASS, CHARGED PION MASSES*	TYPICAL MODE OF CREATION†	LIFETIME, SECONDS	DECAY	DECAY PRODUCTS†
ω^0	omega	not yet observed	5.60	$p + p \rightarrow \omega^0 + 2\pi$	$\sim 10^{-23}$	strong	3π
$\rho\,(-, 0, +)$	rho	not yet observed	5.42	$\pi + p \rightarrow \rho + p$	$\sim 10^{-23}$	strong	2π
η	eta	not yet observed	3.93	$K^- + p \rightarrow \eta^0 + \Lambda^0$	$\sim 10^{-23}$	strong	3π
K^0	kaon	$\overline{K}^0$	3.57	$\pi^- + p \rightarrow K^0 + \Lambda^0$	1×10^{-10}, 6×10^{-8}‡	weak	2π; 3π, etc.
K^-	kaon	K^+	3.54	$\pi^+ + n \rightarrow K^+ + \Lambda^0$	1×10^{-8}	weak	$2\pi, \mu^- + \bar{\nu}_\mu$, etc
π^-	pion	π^+	1.00	$n + p \rightarrow \pi + 2p$	3×10^{-8}	weak	$\mu^- + \bar{\nu}_\mu,\; e + \bar{\nu}_e,$
π^0	pion	π^0	0.97	$p + p \rightarrow \pi^0 + 2p$	2×10^{-16}	electromagnetic	$\pi^0 + e + \bar{\nu}_e$; 2γ

* Charged pion mass = 273 electron masses.

† Where charges are not specified, any charges that add up to the right amount are possible.

‡ K^0 is unique in having two kinds of decay and two lifetimes.

TABLE 7-4. LEPTONS

PARTICLE AND CHARGE	NAME	ANTIPARTICLE	REST MASS, ELECTRON MASSES
μ^-	muon	$\bar{\mu}^- = \mu^+$	207
$e\,(-)$	electron	$\bar{e} = e^+$ (positron)	1
ν_μ	neutrino	$\bar{\nu}_\mu$	0
ν_e	neutrino	$\bar{\nu}_e$	0

PARTICLE AND CHARGE	TYPICAL MODE OF CREATION	LIFETIME, SECONDS	DECAY	DECAY PRODUCTS
$\bar{\mu}^-$	$\pi^- \rightarrow \mu^- + \bar{\nu}_\mu$	2×10^{-6}	weak	$\mu^- \rightarrow e + \nu_\mu + \bar{\nu}_e$
$e\,(-)$	$\gamma + p \rightarrow e^+ + e + p$		stable	
ν_μ	$\pi^+ \rightarrow \mu^+ + \nu_\mu$		stable	
ν_e	$\mu^+ \rightarrow e^+ + \bar{\nu}_\mu + \nu_e$		stable	

TABLE 7-5. NUCLEONS AND THEIR RESONANCES

RELATIVE MASS	CHARGES	RELATIVE MASS	CHARGES
1	$0, +1$	1.79	$0, +1$
1.32	$-1, 0, +1, +2$	2.04	$-1, 0, +1, +2$
1.5	$-1, 0, +1, +2$	2.33	$0, +1$
1.62	$0, +1$	2.51	$-1, 0, +1, +2$

There is another particle outside the classification—the still hypothetical graviton, associated with the gravitational field and presumably neutral and massless.

At first, one is inclined to look at the figures for the lifetimes and conclude that all unstable particles except the neutron are extremely unstable. But there is actually a great range in these short lifetimes. The shortest belong to the resonances, which decay in a length of time about equal to that required for light to cross a neutron or a proton. Since no signal can travel faster than light, it is difficult to imagine a particle disorganizing itself any more quickly than that. In comparison, the particles that live for 10^{-10} second are very stable, for the ratio of the two times is 1 second to a million years. It seems therefore that we have to understand the stability of the mesons and hyperons rather than their instability.

Another difference points up this distinction. When fields interact, as they do in the creation and destruction of a particle, they interact with a certain characteristic strength. If we know how a particle is produced, we can calculate the strength of the interaction between the fields involved, and we can then estimate how long it will take the particle to decay. With the observed rates of productions of strange particles, these calculations give decay times of about 10^{-23} second. This value corresponds to the decay of the resonances but not of the other particles.

The modern explanation of the difference in lifetimes is that particles will decay in about 10^{-23} second if they can, but that some cannot, and they wait for a very much longer time until an entirely different mechanism carries them off. The interaction by which particles are created and the fastest decays occur is called *strong,* and that which takes over if a particle cannot decay strongly is called *weak,* meaning that it is some 10^{-13} times as strong as the other. Presumably, the weak interaction is always present, so that particles can be created weakly, too, but it is unlikely that we shall ever observe such an event, since about 10^{13} times as many particles would be created at the same time by strong interactions. Even weaker than the weak interactions (and, in fact, about 10^{-13} times as strong again) are *gravitational* interactions. We do not know or know how to find out whether these play any part in the physics of elementary particles. One might guess that just as most particles that are stable with respect to strong decay will at length decay weakly, so protons and nuclei, which are stable with respect to weak decay, will someday disintegrate gravitationally. This does not appear to be so, however, for 10^{13} times 10^{-10} second is only 10^3 seconds, or about a quarter of an hour, and the particles in question are stable vastly longer than that. (The neutron happens to live about this long, but we know that it decays via a weak interaction, and we know why it is so slow about it.)

If, as now seems possible, all weak interactions are mediated by the intermediate particle W, described in Chapter 6, the coupling strength of W is only half as weak as that of the interactions it produces, and its lifetime is correspondingly about midway between those of the strongly and the weakly decaying particles.

The tables show several regularities. Leptons and baryons other than W obey the exclusion principle; mesons do not. Baryons obey a conservation law stating that the number of baryons minus the number of antibaryons is unchanged in any interaction. There are two groups of leptons, e, ν_e and μ, ν_μ, and each group obeys the analogous conservation law, as can be seen in the two typical reactions,

$$\pi^- \rightarrow \mu^- + \bar{\nu}_\mu \qquad \mu^- \rightarrow e + \nu_\mu + \bar{\nu}_e$$

Mesons do not obey a conservation law and may be created in any numbers if the energy is available. The tables present the processes by which the strange particles form and some of the processes by which they decay.

It is impossible to give here any complete account of the experimental methods that have been used to find the recent elementary particles and measure their properties. Three of the most important, however, have resulted in the identification of the resonances at Berkeley, the discovery of the second kind of neutrino at Brookhaven, and the search for the W at Geneva.

Resonances

In the last few years it has become possible to detect particles that decay by strong interactions. It is known that a negative kaon (K^-) shot against a proton will sometimes produce a lambda particle (Λ) and a pair of oppositely charged pions. Figure 7–7a shows essentially what happens. But, according to Figure 7–7b, there may also be formed some very short-lived intermediate particle, which subsequently decays into a lambda particle and a pion. No single bubble-chamber photograph would confirm the existence of this intermediate particle, which we shall designate by the letter Y, because its path through the chamber would be too short to observe. From a dynamic point of view, however, the two processes are

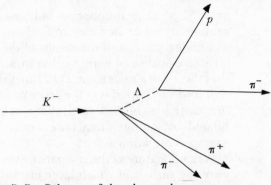

FIGURE 7–7a Scheme of the observed process
$$K^- + p \to \Lambda + \pi^+ + \pi^-$$
$$\hookrightarrow p + \pi^-$$

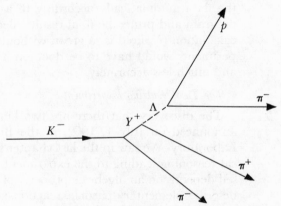

FIGURE 7–7b Scheme of the process in Figure 7–7a as it might take place via the intermediate short-lived particle Y according to the reaction
$$K^- + p \to Y^+ + \pi^-$$
$$\hookrightarrow \Lambda + \pi^+$$
$$\hookrightarrow p + \pi^-$$

quite different, for in one case the collision between kaon and proton produces three particles and in the other it produces only two, and the statistical distributions of angles and energies differ greatly in the two cases. Since this is so, starting in 1960 a large number of bubble-chamber photographs were analyzed at Berkeley to find out which decay process was taking place, and the results of the analysis gave unmistakable evidence of both mechanisms. They showed further that the Y has a mass some 2710 times that of an electron and that its decay period is of the order of 10^{-23} second. It was the first particle to be discovered that was both formed and annihilated by strong interactions, and now about fifty other examples are known.

A novel feature of the resonance experiments was that they were the earliest in which large electronic computers played an essential part. The machines that scan and measure bubble-chamber photographs deliver their results on punched cards. These are fed directly into a computer, which analyzes the data automatically according to a previously designed program and prints the final results. Because the amount of calculation required is so great, without computers such experiments would have to be done on a much smaller scale and much less accurately.

The Two-Neutrino Experiment

The discovery that there are two kinds of neutrinos was announced on July 1, 1962, at the Brookhaven National Laboratory. We saw in the last chapter some of the theoretical reasoning leading to the experiment. The experiment itself depends on an algebraic property of the expressions that describe elementary-particle reactions: a particle may be moved from one side of an expression to the other provided that it is replaced by its antiparticle. Thus decay of a neutron,

$$n \rightarrow p + e + \bar{\nu}$$

implies that

$$\nu + n \rightarrow p + e$$

and also that

$$\bar{\nu} + p \rightarrow n + e^{+}$$

Let us look at two other pairs of decay reactions,

$$\pi^{+} \rightarrow \mu^{+} + \nu \qquad \pi^{-} \rightarrow \mu^{-} + \bar{\nu}$$

and

$$K^+ \to \mu^+ + \nu \qquad K^- \to \mu^- + \bar{\nu}$$

in which neutrinos are produced together with muons. If there were only one kind of neutrino, then we would expect that the neutrinos and antineutrinos from these four decay processes could be used to trigger the two preceding ones and produce electrons and positrons that could be observed. The great difficulty of the experiment lies in the fact that we are here dealing with weak interactions, and the probability of any process involving them is relatively low. It is enhanced at higher energies, and hence the Alternating Gradient Synchrotron was used (though not, for technical reasons, at its full energy) to produce high-energy muons and kaons from which came high-energy neutrinos and antineutrinos. In the course of 800 hours of operation, some fifty events caused by neutrinos were observed as they occurred in a large spark chamber that was triggered by counters. None of the neutrinos interacting with protons in aluminum plates in the chamber gave rise to electrons or positrons. Instead, the reactions

$$\nu + n \to p + \mu^- \qquad \nu + p \to n + \mu^+$$

took place. Thus neutrinos created along with muons seem to be unable to create anything but muons. It is now necessary to show, though the experiment is even more difficult, that neutrinos born in processes involving electrons will produce electrons and not muons. Any other result would be a great surprise. The Brookhaven experiment seems to have provided us with the first new stable particle since Wolfgang Pauli first talked about neutrinos thirty years ago.

The W Experiment

The high-energy facilities at Geneva were employed for 2 months of 1963 in the search for the intermediate particle called W. Like the experiment just described, this one centered around the impact of high-energy neutrinos on nuclear particles. When ν_μ's from the decay of energetic pions and kaons strike protons in the target, the theory of W interactions predicts the process

$$\nu_\mu + p \to W^+ + \mu^- + p$$

after which, in about 10^{-17} second, the W's should decay into muons or positrons,

$$W^+ \to \mu^+ + \nu_\mu \quad \text{or} \quad W^+ = e^+ + \nu_e$$

with about equal probabilities. The central experimental problem is to detect the pairs of muons that accompany the birth and disintegration of a W, and the difficulty is to distinguish them from muon-pion pairs arising from the competing reaction,

$$\nu_\mu + p \to \mu^- + \pi^+ + p$$

which occurs at the same time. This distinction can be made because it is known that, on the average, pions travel less far in a spark chamber than muons, since they interact more strongly with any nuclei they encounter. Therefore, if enough events are observed, one can establish statistically whether all of them may reasonably be ascribed to muon-pion pairs or whether the desired pairs of muons may also have been involved.

At a conference on elementary particles in Siena in October, 1963, it was announced that in several thousand hours of operation seventy-six pairs of particles had been observed and that probably only about one-third had included a pion. This preliminary result was received with cautious optimism, but the final results, announced in New York in February, 1964, were not conclusive, and corroborative experiments are in progress at several laboratories.

INTERPRETATIONS

In trying to construct a theory of elementary particles with the mass of data before us, we are in much the same position as the physicists before quantum mechanics who tried to understand the structure of atoms from a study of their spectra. It must have seemed to some of them as if they could never succeed. Finally they realized that the facts essential to an understanding of atoms were not all in the tables of spectra; many are in our knowledge of ordinary physics, for the atom is visualized as a mechanism in which charged particles interact according to the laws of electricity and move according to the laws of motion. To be sure, these laws have had to be somewhat modified, but it is perfectly possible for us to understand atoms, at least roughly, in terms of Rutherford's model of electrons circling around a nucleus.

What concerns us now is that for the elementary particles, or at least for some of them, there may be no such model. We may have passed beyond the stage at which anything in our ordinary experience can help us. To the extent that this is so, our construction of a theory will have to be a purely mathematical act of creation rather than an application of familiar ideas to new ends.

The first task is to examine the data and see what clues they afford us. There are several. There is, for example, a neutral meson resonance called eta (η), which decays quickly into three pions. Its mass is 548.5 ± 0.6 Mev.* On the other hand, the total mass of the four pions $\pi^- + \pi^+ + 2\pi^0$ is 549.2 $\pm$ 0.1 Mev. Is an eta just four pions bound together with a binding energy in the neighborhood of 1 Mev?

---

*The ± 0.6 indicates the uncertainty of the experimental measurement; the mass of the eta is probably between 547.9 and 549.1 Mev. The masses of the particles in this section are expressed in energy units (Mev) for the following reason. Suppose that a pion disintegrates into a muon and a neutrino. The pion's mass is 2.448×10^{-25} gram; transformed to energy units by the relation $E = mc^2$, this becomes 139.6 Mev. Similarly, the muon's mass is 105.6 Mev. When a pion decays, the difference, or 34 Mev, seems to have disappeared. It has not, however, for 34 Mev is exactly the total energy of motion of the muon and the neutrino as they start out. The mass of a disintegrating system is always greater than the total mass of the pieces into which it disintegrates. Consider on the other hand a compound system that is perfectly stable, such as a typical atom. In order to separate it into its fragments, work must be done, that is, energy must be added, and energy has mass. Thus in this case the total mass of the separate fragments is greater than that of the original system. A helium nucleus, for example, has a mass that, expressed in energy units, is 28.3 Mev less than that of the two neutrons and two protons composing it. Therefore, this is exactly the energy that must be added if one wishes to take it apart. It is called the binding energy.

TABLE 7–6

PARTICLE	OBSERVED CHARGES	MASS, MEV	DIFFERENCE, MEV
Ξ^*	$-1, 0$	1533 ± 3	
Y_1^*	$-1, 0, +1$	1385 ± 5	148 ± 6
$\mathcal{N}_1$	$-1, 0, +1, +2$	1237	148 ± 5

An interesting form of regularity is found in equal spacings between mass values. Including the different electric charges, we can pick out the set of resonances given in Table 7–6, where $\mathcal{N}_1$ is the first nucleon resonance in Table 7–5.

Equal spacings are common among atomic energy states, and there we understand the situation. These are states that would have exactly the same energy if the atom and its environment showed perfect spherical symmetry. A slight disturbance of the symmetry separates the levels slightly. It is possible that a similar thing happens here, but we know that the symmetry here is not the ordinary symmetry of spatial directions. Rather, it is some sort of abstract mathematical symmetry. If we knew all about the mechanics of the elementary particles, it would probably be easy to tell what it is that is symmetrical; since we know nothing, we can do little more than run through the list of reasonably simple types of symmetry known to mathematicians and see whether any of them seem to explain the observed groupings of levels. There is at present one leading candidate, discovered independently by Murray Gell-Mann of the California Institute of Technology and Yuval Ne'eman, an Israeli research student in London. Precisely because the conclusions do not yet spring from any specific model of elementary particles, the sense of the theory is hard to convey, but a few definite statements can be made even at this early stage.

The new theory, called the Eightfold Way, predicts for example that the masses of the nucleon, xi, lambda, and sigma will be connected by the formula

$$\tfrac{1}{2}(m_N + m_\Xi) = \tfrac{1}{4}(3m_\Lambda + m_\Sigma)$$

The measured values of the left and right sides are 1129 and 1134 Mev, respectively, and there are two other relations that come out equally well. On the other hand, the theory predicts that Table 7–6 should be completed by entering at the top the single particle called omega (Ω), with a charge of -1 only and a mass of $1533 + 148 = 1681$ Mev.

Today's theory starts out to answer the question, "If there are so many ways for particles to decay, why are not all ways possible?" For example, why have the neutrons and protons composing most of our mass not broken down into wandering swarms of electrons, positrons, photons, and neutrinos? There must be basic principles of physics that make some disintegrations impossible, and we think we know what they are: the laws of conservation.

It appears that in any process of creation or decay, the total energy (counting mass as energy) is the same after the process as before. We can say this fairly confidently *even though we do not understand how the decay takes place.* Similarly, there is a conserved quantity called angular momentum, which describes (roughly) the spin of the entire system, and we suppose that there are others. Why do we suppose that these quantities remain constant if we do not understand what is going on? To this we can give a definite answer: the conservation laws arise out of symmetry. Angular momentum, for example, is conserved because to an elementary particle all directions of space are equivalent, and if it is already spinning around one axis at a certain rate, there is nothing in its environment that can cause it to change. (To us, of course, all directions are not equivalent. The direction called *down* is singled out by the earth's gravitational field, but gravitational forces are apparently too weak to affect elementary particles sensibly.) The conservation of energy arises, perhaps surprisingly, from a symmetry in time. One can be sure that if he begins a certain experiment at 11 o'clock, the result will be exactly the same as if he had begun it at 9 o'clock or at any other time.

The conservation laws give rise to a pattern of forbidden processes among all the others that are allowed, and nature seems to follow what Gell-Mann has called the Principle of

Pure Totalitarian Government: "Everything that is not forbidden is compulsory." We can easily verify that energy and angular momentum are conserved when a particle is created or decays, but these are not nearly all. For example, the number B, defined as the number of baryons minus the number of antibaryons in any reaction, is always conserved; and the tables show that it is exactly this law that keeps us from disintegrating. We are in a curious situation. We know the conservation laws, but we do not know their underlying dynamic basis; that is, we do not know the kind of symmetries responsible for them.

Let us assume, however, that there are such symmetries. The theory of symmetries (mathematicians call it the theory of groups) is highly developed, and since the doctoral thesis of Élie Cartan in 1894, we have been aware of all the kinds of symmetry possible. Each kind should lead to a certain pattern of conservation laws, and by looking at the conservations shown in the behavior of elementary particles, we may hope to find what symmetries are involved in them, even if we do not know enough about the internal dynamics of the particles to recognize what is symmetrical.

In 1961, when relatively few particles were known, Gell-Mann and Ne'eman independently put forward a type of symmetry known to mathematicians as SU(3), and as more particles were discovered, they were fitted into this pattern. SU(3) is a set of eight symmetries, and therefore Gell-Mann has called his theory the Eightfold Way, after the eight basic precepts of the Buddha, which lead to the cessation of pain.

We have mentioned earlier the particle known as omega, which would fill out a set consisting of four N_1's, three $Y_1{}^*$'s, and two Ξ^*'s. According to the eightfold way, the omega *has* to exist, and consequently there was great rejoicing when on February 23, 1964, a paper appeared, signed by thirty-three staff members of the Brookhaven National Laboratory, which stated that in a series of 100,000 bubble-chamber photographs taken at the Alternating Gradient Synchrotron, a single event had been observed that could with reasonable confidence be called an omega. The particle was formed in the collision of a negative kaon with a proton, and the reaction that probably followed is a compendium of those that we have studied.

$$K^- + p \to \Omega^- + K^+ + K^0$$
$$\hookrightarrow \Xi^0 + \pi^-$$
$$\hookrightarrow \Lambda^0 + \pi^0$$
$$\hookrightarrow p + \pi^-$$

The pions and kaons undergo further decays not written down. The tracks of all the charged particles, plus those due to the charged decay products of the neutral pion, have been identified and measured, and all are consistent with an omega having a negative charge, a mass equal to 1686 Mev (as against about 1681 Mev predicted above), and a lifetime of about 10^{-10} second.

No theory can be proved by a single experiment, much less by a single series of photographs; a theory only becomes more and more certain as more and more bits of independent evidence accumulate. But the Brookhaven photographs, rich with promise, suggest that we are on a way that may indeed lead to the cessation of pain.

DECLINE OF FIELD THEORY

Earlier in this book an attempt was made to show how, during the last century, ideas about fields have superseded ideas about mechanisms for explaining the natural world. But this does not mean that the ultimate explanation of everything is going to be in terms of fields, and indeed there are signs that the whole development of field theory may be nearer its end than its beginning. These signs are most obvious in the theory of strong interactions, though they can be seen in almost all parts of fundamental theory.

The subject of strong interactions began in 1935 with Yukawa's brilliant guess that the strong force holding a nucleus together is a manifestation of a quantized field whose particles, now called pions, should be about 200 times as massive as electrons. The guess was confirmed, but it should be remarked that Yukawa's estimate of the mass of a pion did not depend on getting the details of the nuclear interaction right. In fact, he did not know how to do the calculation, and, moreover, in thirty years of intense effort on an international scale, no one else has been able to do it. The only successful analyses of strongly interacting systems have been those that somehow avoid the main difficulties and obtain their answers by tricks. Thirty years is a long time in modern science, and a physicist can perhaps be pardoned if he wonders whether it will ever be possible to solve the equations of the theory of strong interactions, or whether they have solutions at all, or whether they even make sense.

In other branches of field theory there are also problems. The theory of electromagnetic interactions, for example, as needed to explain the outer structure of an atom, has been phenomenally successful, but the success has been attained by the use of stratagems for getting around basic mathematical difficulties that have been impossible to surmount. This situation would perhaps be bearable if it were not for the shipwreck of Yukawa's theory. Recently Geoffrey Chew of the University of California at Berkeley wrote, "I do not wish to assert that the conventional field theory is necessarily wrong, but only that it is sterile with respect to strong interactions and that, like an old soldier, it is destined not to die but just to fade away." What is the alternative?

Let us notice two things about field theory. First, although it assumes that everything we observe can be explained in terms of fields, the fields themselves are never directly perceived by our senses and so are, at least to some extent, abstractions. That is, fields do not necessarily have to exist. Second, let us remember that Newton's greatest contribution to physics was his insistence that we must try to understand physical occurrences through certain fundamental laws of motion, which tell us how a physical system changes as time passes. Newton gave us the laws of motion of material objects, Maxwell those of the electromagnetic field, and Schrödinger and Dirac those of the matter field. But there are many things in nature whose behavior in time is not their most obvious feature: the mass of an elementary particle, for example, or the intensity of a beam of particles scattered in a certain direction. In recent years a new kind of physical theory has been developing, based on a series of papers by Heisenberg in the early 1940's, which goes back to the ancient idea of physical explanation in terms of how things are rather than how they change. This theory involves neither the concept of a field nor equations of motion, though one can, if one wishes, use it to express the conventional field theory and its shortcomings.

Any physical theory is like a sausage grinder into which one puts observed facts about the world, together with one's physical notions as to what is going on. What comes out is other facts about the world, which can be tested by experiment. The new sausage grinder (called the *theory of the S matrix,* or the *dispersion theory*) is different from the old in that one puts into it facts but very few physical notions (no fields, no equations of motion) and only certain general but definite hypotheses of a mathematical nature. The machine is new, and the crank turns stiffly. Nobody knows how to solve the purely mathematical problems it produces. But there has been an impressive amount of success, especially in the realm of strong interactions where the field theory is most helpless. Some physicists, notably Chew and his followers, proclaim the *S*-matrix theory as the physics of the future. Others, more cautious, seek to reconcile it with a subtler form of field theory. In any event, at present, for the purely practical purpose of grinding out answers, it seems our best hope.

STRUCTURE OF NUCLEAR PARTICLES

The properties of elementary particles that we have mentioned so far reflect chiefly the particles' dynamic behavior. It is natural to ask a simple question, "What would a particle look like if we could see it?" The way to answer this is to build a microscope and find out, and this has been done. The resolving power of such a microscope must be very great, and we immediately run into the basic limitation of any such instrument, which is the wavelength of the radiation it uses. This affects any attempt to resolve the structure of something by directing a wave at it. An optical microscope can be employed to observe objects down to the scale of a few wavelengths of light, say 10^{-4} cm, but at this point the image gets fuzzy, and no combination of lenses will clear it up. It is possible to improve the optical microscope a little by using blue or violet light, which has a slightly shorter wavelength than the other colors, and for a few years there were "ultramicroscopes" using ultraviolet light. These were made obsolete during the 1930's by the electron microscope, which uses a beam of electrons in much the same way that an optical microscope uses a beam of light, except that it requires electric and magnetic fields for focusing instead of lenses. We have already seen that an electron has a wavelength, and this imposes the same basic limitation as before, but the wavelength is much shorter, and the instrument will show correspondingly smaller objects.

It is a fundamental property of matter waves that as the wavelength decreases, the energy of the particles increases. This raises difficulties in biological studies, for high-speed electrons tend to damage a specimen, but the difficulties can be partly overcome by technical tricks, and magnifications of over a million have been achieved. To go further, one can either abandon the use of electrons or push them to still higher energies.

Suppose, for example, that we wish to study the arrangement of atoms in a crystal. To use energetic electrons is impossible because they tear the crystal surface to pieces. The only way of attaining the necessary wavelength of about 10^{-8} cm is by means of particles, much heavier than electrons, whose wavelengths are short but whose energies are not so high. Erwin Müller of Pennsylvania State University has constructed a device in which helium atoms acquire electric charge next to a fine metal point and are accelerated away from it by an electric field to project images of the individual atoms of the point on a screen an inch or so away. Figure 7–8 is one of Müller's pictures of a tungsten tip. Each white dot is an atom, and the complex pattern is what one expects from looking at the rounded surface of a regular crystal. Müller's device is not yet very flexible in its applications, but it is ex-

FIGURE 7–8 Arrangement of atoms at the tip of a fine tungsten needle, as shown by the field-ion microscope ($\times$ 1,400,000). (From E. Müller.)

cellent to use on metal surfaces and on people who say, "Why do you talk so much about atoms if you have never seen one?"

The other way to observe smaller and smaller things is by means of electrons of higher and higher energies. The apparatus for this procedure looks less like a microscope and more like a high-energy accelerator with complex equipment for analyzing the beams of electrons that emerge from the sample under observation. Electrons continue to be used because they interact with target particles in a manner that is relatively well understood (this is, of course, essential if the data are to be correctly interpreted), and the objects studied are neutrons, protons, and nuclei, which are not damaged by the electrons. The techniques involve the use of electrons in the billion-volt range and were developed largely at Stanford University by Robert Hofstadter, who received the Nobel Prize for his work in 1961. Figure 7–9, drawn a little out of

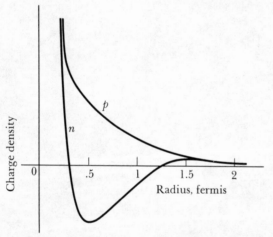

FIGURE 7–9   Variation in charge density of a neutron or proton with distance from the center of an atom. (One fermi = 10^{-13} cm.)

scale in order to emphasize its salient features, shows how the densities of electric charges of protons and neutrons are found to vary with distance from the center. Evidently the neutron is only neutral as a whole and contains shells of both positive and negative charges. The particles do not have sharp boundaries but trail off to nothing with increasing distance. The principal center for this research at present is the Cambridge Electron Accelerator, a joint project of Harvard University and the Massachusetts Institute of Technology. The machine produces electrons with energies up to 6 Gev; these particles move so rapidly that if they were to race a beam of light to the moon, they would lose by only 5 feet. For some time it was thought that a nucleon consisted of a cloud of charge surrounding an impenetrable core, but as electrons of higher energies are used, a nucleon becomes increasingly transparent, and the data now suggest that perhaps there is no core at all.

If nothing else, this chapter should have taught us to doubt whether the so-called elementary particles are really very elementary. From the beginning the S-matrix theory of strong interactions has given rise to formulas to describe interacting particles that were much more complicated than anyone expected, and it is at length understood that this is because each particle so described has a complex structure. In a sense, each particle acts as though it were a composite of all the others with which it interacts strongly. This composite may be so tightly bound that it loses all resemblance to the particles that compose it or so loosely bound that it becomes more like a nucleus, in which the particles keep their individuality while remaining close together in space. The theory can be checked in detail by the electron-scattering experiment just discussed. The theorists who first attempted such a check found in 1959 and 1960 that they could get agreement with experiment only if they assumed that there exist both two-pion and three-pion states in which the pions are strongly attracted together, though not bound permanently. The search that led to the discovery of the resonances called rho ($\rho \rightarrow \pi + \pi$) and omega ($\omega \rightarrow \pi + \pi + \pi$) was inspired by their calculations, and the theory now explains all the general features of Figure 7–9. As concerns both insight and practical results, we have learned more about strong interactions since 1960 than in all the previous years since Yukawa's invention.

CHAPTER *8*

Cooperative Phenomena

It would be nice to think that once we understand the nature of the elementary particles and their interactions we will know all about physics. Unfortunately, there is more to understanding how the universe behaves than understanding how its basic entities behave individually. One can know a friend quite intimately and still be astonished by his conduct at a noisy party; there are aspects of the behavior of a person or of an atom interacting with a large number of others that are not suspected from the behavior of the individual alone.

The reasons for this in physics are twofold. First, some properties of a physical object simply do not appear from a study of it in isolation from its fellows. Second, the mathematical problem of calculating from a knowledge of the behavior of one member of a species what a large number of them will do when put close together in interaction is so complicated that in many cases we do not know how to approach it at all. It is therefore possible to be surprised by things that we discover happening in closely interacting physical systems merely because our analytical technique is too primitive to lead us to expect them. Examples lie close at hand. Physicists think that they know all there is to know about the foundations of chemistry, but the average physicist is not a chemist at all, for

131

chemical compounds, viewed in terms of the atoms that compose them, are structures so elaborate that their properties are usually established by experience in the laboratory before they can be discussed from first principles. As chemistry is to physics, so biology is to chemistry, and accordingly it is even more difficult to explain a biological process in terms of the atoms taking part in it. But the difficulty must not obscure the fact that such an explanation is in principle possible today and that in a century or so it may, on a limited scale, be an actuality.

At present, to approach chemistry through physics is not, except at an advanced level, very enlightening. Therefore, a few examples taken from physics of what are called *cooperative phenomena* will be given here to show how an understanding of them deepens our knowledge of nature and also to show something of the difficulties involved in achieving this understanding.

In the first place, let us look at some phenomena that typically depend on the presence of a large number of particles and are not found if there are only one or two. Suppose, for instance, that we have an airtight chamber with gas molecules in it. If the molecules are numerous, any part of the walls of the chamber will experience a steady rain of tiny impacts as molecules bounce against it. (It may help us to visualize the situation to know that the molecules of a gas under ordinary conditions move at some hundreds of miles an hour and suffer a collision every 10^{-5} cm or so.) The impacts on the walls are too feeble and too frequent to be detected separately except by special means; yet they have the effect of a steady outward force—the force that, for example, keeps a tire inflated.

One of the properties of an elastic medium such as a gas is that it conducts waves of alternating compression and rarefaction, which, if they are in the range of frequencies to which our ears respond, are called sound. Sound waves are easy to imagine if one thinks of a gas as a uniform, smoothed-out medium, but if one things of a gas in terms of individual molecular motions, the description of sound becomes complex.

Suppose that the chamber contains only three gas molecules. These molecules bounce around hitting the walls of the chamber and very, very occasionally colliding in midair. What then does it mean to talk about sound in such a chamber? Sound is a wave of pressure exerted by the collisions of the molecules; since the molecules rarely collide, the chamber containing only a few molecules cannot be said to contain sound at all. There have to be millions of molecules in the chamber, with collisions occurring all the time, before one can refer to anything like the propagation of a wave through it. Sound is thus a cooperative process.

As a second example, suppose that the molecules in the chamber were molecules of water vapor and that we were to lower the temperature of the whole system. The molecules would move more and more slowly, and because they tend to attract each other at short distances, there would come a time when they would be moving so slowly that if they happened to collide they would stick and go off together. This sticking together is what is involved in the condensation of a vapor into a liquid. But although it explains condensation, in one sense, it does not explain the whole process. Sticking together does not explain, for example, why water vapor at atmospheric pressure always condenses into water droplets at a temperature of exactly 100°C. More is taking place than the random association of a pair of molecules that have slowed down somewhat. The condensation setting in throughout a cloud of water vapor at a temperature of exactly 100°C is something that is typical of the cooperative interactions in large numbers of its molecules and that cannot be predicted from studying a few of them.

As another example, there are many properties of matter that depend specifically on the existence of a crystalline state. A crystal is a sample of matter in which, at least over small regions, the atoms are arranged in some regular pattern, like the bricks of a wall. Crystals have a number of characteristic properties, such as the way they transmit light, that samples of the same material in noncrystalline states do not have. Clearly, it is not possible to talk about a crystal until one has a large number of atoms arranged appropriately, any more than it is possible to talk about masonry in reference to only three bricks.

The most difficult and important problems in cooperative physics have to do with dynamics. Let us consider a very simple system whose dynamic behavior is well known. A violin string is composed of a very large number of molecules of protein. Each of these molecules moves back and forth when the string is bowed. But this does not give us a very useful picture of what happens. It is much better to talk about the behavior of the string as a whole than that of its individual molecules.

Suppose that for simplicity we consider a string composed of only three molecules. Let these three molecules be like three beads attached by a weightless thread. Let us see how many different degrees of freedom this system has. By degrees of freedom we mean something very definite, namely, the number of numbers that are necessary to specify precisely where the three beads are. If the beads can only move up or down, as in Figure 8–1, the position of each bead can be specified by its distance above or below the horizontal line. Since there are three beads, there are three degrees of freedom for this complete system. Now let us look at the entire string.

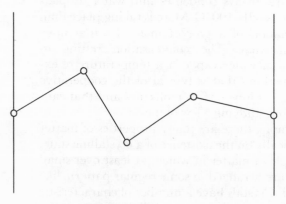

FIGURE 8–1   Typical arrangement of three beads, each free to move up and down, on a string.

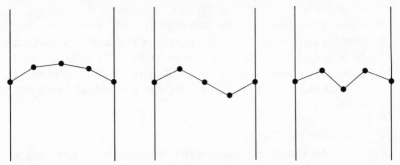

FIGURE 8–2   Three types of motion that are repeated
again and again if all three beads oscillate up and down with the same frequency.

There are three simple and symmetrical ways in which this
string could vibrate as a whole if it were plucked or bowed.
Figure 8–2 shows them. They have successively higher frequencies, and no one could pluck or bow the string to make it
give forth any pure tone that would be higher than the highest of them. This illustrates the fact that the number of degrees of freedom of a cooperative system is the same however
we count them (as long as we count correctly!). But in some
cases it is more meaningful to count in one way than in another. In this particular case, if we are concerned with the
behavior of the entire string with the three beads on it, it is obviously more interesting that it can produce three different
notes than that there are three different numbers giving the
positions of the three beads individually. We can summarize
by saying that the number of *modes of motion* of a physical system equals the numbers of its degrees of freedom.

The real violin string has something like 10^{22} molecules of
protein in it, and therefore it has some 10^{22} modes of motion.
A few of these are heard as different notes, or harmonics, when
a violin string is played—maybe half a dozen; in some other
musical instruments the number of harmonics heard is perhaps as many as twenty. But the number that one hears is not
nearly so large as the number of possible harmonics, because

most of the possible characteristic motions of the string as a whole take place at such exceedingly high frequencies that there is no way to excite them; in any event, one would not be able to hear them if they were excited. In fact, the overwhelming majority of the modes of motion of the violin string is such that in trying to excite them one would break the string.

NUCLEAR STRUCTURE

Let us examine a cooperative system that is receiving a great deal of attention nowadays—the nucleus of an atom. This, as we saw in Chapter 2, is a tightly packed mass of neutrons and protons containing roughly equal numbers of each. A typical nucleus of medium weight might have a hundred particles in it. Each particle has three degrees of freedom, unlike the simple example given earlier, for it can move east and west, north and south, and up and down; therefore, one can specify the position of any one particle by giving three numbers. Thus the nucleus of a hundred particles has three hundred degrees of freedom. There are three hundred possible modes of collective motion, and each gives rise to a whole family of different energy levels analogous to those of an atom. In principle, then, a nucleus has a very large number of possible energy levels. Fortunately, most of them are not observed for the same reason that they are not observed in a violin string: to excite the particular modes of oscillation corresponding to them would involve the exertion of forces on the nucleus that would destroy it.

Nuclear energy levels are the basis of a branch of nuclear theory known as the theory of the nuclear shell model. This theory is now highly developed but does not explain more than a certain number of the known levels. A nuclear energy level diagram is quite complicated, as is obvious from Figure 8–3, but it begins to appear that many, at least, of the energy levels in this diagram arise from relatively simple modes of motion. For example, all the particles in a nucleus except one may act as a relatively inert core while this one particle moves around and through the core like the electron in a hydrogen

Energy, Mev

} 28 levels

FIGURE 8–3   Energy levels of the nucleus of neon-20 (10 protons and 10 neutrons). The scale is roughly one millionth that of Figure 2–4. In the crowded regions the levels are too close together to show in detail. The top of the diagram does not represent the limit of the levels but merely the limit of investigation.

atom. Thus different modes of excitation of the nucleus will merely give different energies to this one particle without affecting the rest. This means that if one can understand the orbits and energy levels of just one particle, he can understand quite a number of nuclear energy levels.

Another simple motion that a nucleus can execute is to rotate as a whole, just as the earth rotates daily. This motion produces simple recognizable patterns of energy levels. In addition, the nucleus can vibrate, elongating and shortening periodically so as to produce a number of vibrational energy levels. Finally, it can have surface waves, which, if one could view them from the side, would look rather like the tides of the oceans on the earth. Looking at the earth, one would see a double wave running around it once every 24

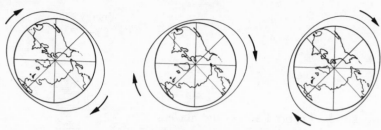

FIGURE 8-4   Exaggerated representation of high and
low tides moving around the earth,
drawn from the earth's point of view.
(From the moon's point of view, the
wave is stationary while the earth turns,
and from the sun's point of view, both
the wave and the earth turn.)

hours, as shown in Figure 8-4. Any one bit of water moves
slowly up and down a few inches, but the wave, which is a
cooperative motion of all the water molecules on the globe,
has a velocity of about 1000 miles per hour. The calculation of the energy levels developing from this kind of cooperative motion is a good deal more difficult than that of
the energy levels mentioned earlier. It has been carried out
largely at the Institute of Theoretical Physics in Copenhagen
by Aage Bohr, the son of Niels Bohr, and several collaborators.

There is still very much to be learned about the structure
of nuclei, and only part of it requires an understanding of the
different motions of a collection of neutrons and protons.
Ideally, we would attack the problem of nuclear structure
like this: First, we would experiment on individual particles,
causing neutrons to collide against neutrons, protons against
protons, and neutrons against protons, and seeing how they
behave in interaction (a great deal of work has in fact been
done in this direction in laboratories all over the world).
From the results of these experiments, we would infer the nature of the forces acting between pairs of nuclear particles.
Armed with this knowledge, we would attack nuclear

theory and try to construct an explanation of nuclear structure from the known properties and interactions of the nuclear particles.

Unfortunately, even if this program offered no insurmountable technical obstacles, there is no reason to believe that we would get the right answer, for we do not know that all the forces acting inside the nucleus are the same as those measured in two-particle interactions outside it. It would, of course, be nice if this were true, but it is also quite possible that the presence of a third or fourth particle within the range of the interaction of the nuclear forces would affect the interaction of any two of them; perhaps we would have to consider three interacting particles, not as A interacting with B, A interacting with C, and B interacting with C, but rather as a total situation involving A, B, and C, in which the interaction between A and B would be quite different if particle C were removed from the cluster.

The possible existence of so-called many-body forces in a nucleus is a seriously complicating factor and is particularly confusing because it is hard to see how one could arrive at any detailed knowledge of such interactions directly. It is easy enough to make one nuclear particle collide with another but very difficult to shoot three of them together so that they interact in any way that can be studied experimentally. Therefore, modern nuclear theory is based on a nuclear interaction partly theoretical and partly empirical, which is chosen so that we can explain some, at least, of the nuclear properties by its use. If it is impossible for us to investigate a many-body interaction by direct experiment, then the question is ultimately a theoretical one: why do nuclear particles interact at all, and can we understand from a fundamental theoretical standpoint the nature of this interaction?

In the last chapter we learned the probable reason behind nuclear interaction, namely, that nuclei interact with a meson field manifested as a field of force. The purely mathematical complexities in a theory of the meson field are so numerous that only recently have any firm results at all been obtained from it, and even these results are not so much definite formulas for what will happen in a given experiment as they are mathematical theorems saying that one measured nuclear property will bear a certain numerical relationship to another. The theory is not able to predict from fundamental considerations what the value of either of these measurements will be.

If we could be confident that within a few years the theory of the meson field and the interaction between nuclear particles could be summarized in neat mathematics it would be reasonable to wait until the meson theory had made definite theoretical predictions about how particles interact and then apply them to a theory of nuclear structure. Next we would use the theory of nuclear structure to calculate the energy levels of the nuclei, and finally we would compare the energy levels so computed with experimental values already known very accurately. This plan has two serious limitations. The first is that it may be many years, even a generation or so, before the theory of nuclear forces yields ripe fruit. Second, the exact calculation of nuclear properties, even supposing that one understood the force laws involved, is a matter of enormous mathematical complexity. Therefore, as we contemplate the future of nuclear theory, it seems to promise a piecing together of small bits of experimental fact and theoretical conclusion into a body of knowledge that will grow slowly rather than sprout overnight.

SUPERCONDUCTIVITY

A metal is a perfect example of a cooperative system. Its basic underlying form is crystalline. The atoms of a metal, unless it has been worked over vigorously, tend to line up into crystalline patterns at least in small regions. Among these atoms wander electrons that are not attached to any one atom in particular but belong to the metal as a whole and act as an electric current. (Substances that do not have electrons free to move from one atom to another are not conductors of electricity; they are called insulators.)

The electrons moving through a crystalline structure are not like animals running through a forest; if they were, the laws of electricity in metals would be completely different from what they are. What happens is that the electrons interact continuously and strongly with the stationary atoms forming the crystalline lattice (as though the animals caromed from tree to tree). The word "stationary" is not quite accurate here because these atoms are vibrating with a thermal motion in the crystal; as an electron passes near them, this thermal motion has the effect of either adding or subtracting energy from it and deflecting it from its path. The electron gets bounced around as it goes through the metal, continually gaining and losing energy. The progress of an electric current is thus more like the diffusion of water through soil or some other spongy medium than like the flow of water down a pipe. (Even though electric impulses are rapid, the electrons themselves drift down a wire at a rate that is ordinarily less than a millimeter per second.)

Now let us see what this model of electrons in a metal predicts with regard to the temperature dependence of the electrical resistance of the metal. If the resistance is affected by the thermal vibrations of the atoms in the metal, then one would suppose that as the metal is heated up the thermal vibrations, becoming more violent, would lead to a higher electrical resistance; this is exactly what happens. Conversely, as the metal is cooled down, its resistance should get lower and lower. What happens as absolute zero is approached? One might at first expect that the resistance would also approach zero. But one of the results of quantum theory is that motion never really dies out in the natural world; the primitive view of absolute zero as a temperature so cold that all random molecular motion ceases and stillness prevails is not correct. What actually happens is that the atoms in the crystalline lattice continue to oscillate even at absolute zero and as the temperature is decreased, the resistance seems to be approaching a value which is not zero. But for almost every pure metallic element, and for almost a thousand metallic compounds, there occurs a transition temperature, somewhere below 18 degrees above absolute zero depending on the material, at which the resistance abruptly vanishes. It does not just become small. If an electric current is started in a closed loop of one of these materials, the current will flow around and around without any diminution whatever as long as the loop is kept cold. This phenomenon is called superconductivity. That it sets in at a temperature characteristic of the material is significant, for this means that it is a cooperative phenomenon like the boiling or freezing of a liquid.

Superconductivity was discovered by H. Kamerlingh Onnes at Leiden in 1911, and for a long time, while other natural phenomena gradually came to be understood, it remained a mystery. In 1957 the mystery began to dissolve, largely through the labors of a group of physicists at the University of Illinois. The clue to superconductivity is found in the fact that the substances which resist superconductivity are at ordinary temperatures unusually good conductors of electricity. Since in these substances the thermal vibrations are in no way unusual, it is clear that the electrons are less strongly influenced by the vibrations than they are in ordinary substances. Thus superconductivity must be a cooperative effect involving two sets of objects, the atoms themselves and the conduction electrons. Another piece of evidence in this direction is furnished by the existence of an isotope effect. Two pieces of lead composed of atoms of different atomic weights are chemically and for most purposes electrically indistinguishable. However, they differ in the temperature at which superconductivity sets in, suggesting that the vibrations of the lead nuclei, which of course do depend upon their masses, are in some way connected with the phenomenon.

The Illinois group, consisting of John Bardeen, Leon Cooper, and Robert Schrieffer, succeeded in breaking the problem down into two parts, and the dénouement was unexpectedly simple. They first proved a theorem having to do with the cooperative behavior of electrons as follows: if the net interaction between two electrons is an attractive one, then there will be certain collective modes of motion of the electrons that will have the properties of superconductivity; that is to say, there will be a large-scale, ordered flow with no dissipation of energy tending to slow it down. The second part of the problem was to show why the net interaction between electrons is attractive. If nothing else were involved, the interaction would obviously be one of repulsion, because electrons all have negative electric charges and like charges repel. Hence, in a superconductive system there must be an even stronger force of attraction. The stronger force is the force exerted by the phonon field.

In Chapter 4 phonons were described as a wave field carrying energy and being emitted and absorbed by particles in a crystal lattice. But wave fields are also force fields, as we have several times had occasion to note. In metals, the effect of the phonon interactions is to produce an attractive force between electrons that in certain cases more than overbalances the force of electrostatic repulsion. When will the phonon field be strong? Phonons are oscillations of the lattice as a whole; therefore, if the electrons interact strongly with the lattice, as we have just seen that they do in a superconducting material, the phonon interaction between electrons will be strong. One can estimate from other electric data the strength of the phonon interaction and predict fairly well in some cases what the transition temperature of a metal will be. But all the pure metals whose atoms are magnetic, as well as their compounds, resist this analysis. Some of them show no isotope effect at all; others show it in the wrong degree. Bernd Matthias of the University of California at La Jolla, who has discovered many of these recalcitrant materials, believes that their electrons may interact magnetically rather than through phonons. There is much about the theory of superconductivity that must yet be cleared up, but it is no longer the scientific enigma that it was a few years ago, and the way is now open for an enormous expansion of low-temperature technology and engineering by the incorporation of superconducting materials into the catalogue of substances available for exploitation in technical devices.

One such application is the use of superconducting current carriers to achieve a high magnetic field. Strong magnets are important research tools in many industries, but until now they have been rather expensive to operate. A typical elec-

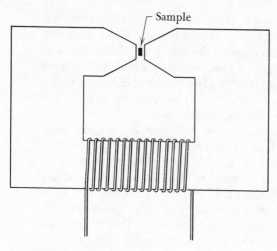

FIGURE 8–5   Electromagnet.

tromagnet is shown schematically in Figure 8–5. It consists of a massive iron yoke with a small gap where the intense field is produced and with hundreds of turns of wire (here only sketched) wrapped around it and fed with electric current in order to magnetize the iron. It is this current that is so troublesome to maintain. A magnet of the largest class, giving a field of 100,000 gauss or more, generally requires over a million watts of power to operate it, which is enough to light a small town, and since all this power is dissipated as heat in the wire windings, a cooling system using thousands of gallons of water per minute is needed as well. Only a negligible amount of the power goes into establishing the magnetic field, and that only at the moment it is turned on. All the rest is wasted in heat, and if the windings had no electrical resistance, they would require no power at all once the field had been set up.

This being so, it seems obvious that if the windings could be made of superconducting material and thermally insulated from the iron yoke so that they could be cooled, a much better magnet would result. But there is a difficulty, for it was discovered early that superconductivity is destroyed by a magnetic field. The wire windings of a magnet experience an intense field when the current flows, and this would cause the magnet to turn itself off. The only hope of getting around this self-defeating property lies in the fact that a current flowing in a very thin layer is relatively impervious to magnetic fields. And in the last decade it has been realized that there are a number of materials, called hard superconductors because most of them are very brittle, in which the current flows in a sponge of narrow filaments along the boundaries of the crystals of which they consist. These materials are mostly compounds of metals; the most important at the moment is Nb_3Sn, a compound of niobium and tin, which can stand a field of 88,000 gauss and has the added feature that it superconducts up to 18 degrees above absolute zero ($18°K$), the highest transition temperature of any substance known.

Several firms are now marketing powerful magnets wound with wires of this brittle and inconvenient material, which consume no more current than that required by the cooling apparatus. And a compound of vanadium and gallium, V_3Ga, has recently been found that tolerates the highest magnetic fields to which anyone has been able to expose it. Within a few years magnets using this material may be functioning at 500,000 gauss, and though nobody expects any dramatically new properties of matter to appear at such field strengths, it will be interesting to see what happens.

LIQUID HELIUM

When liquids are cooled to low enough temperatures, they freeze. These temperatures are high for liquid metals and quite low for most gases. There is a single exception, however—helium gas, which remains liquid, certainly to within one one-hundredth of a degree of absolute zero and probably all the way down. If helium is cooled at ordinary pressures, it condenses at a temperature of 4.2 degrees above absolute zero ($4.2°K$) into a liquid of very low density but otherwise ordinary properties. If it is cooled still further, to below $2.19°K$, which is its transition temperature, this liquid becomes very unusual indeed. All other liquids, for example, have a property known as viscosity. If one tries to pass a liquid down an extremely thin tube or force it through a layer of sand, it will not flow rapidly because of friction from sliding against the solid boundaries and against itself. But if one subjects liquid helium below 2.19°K to the same treatment, it is found to move with virtually no hindrance at all and at a fast rate of speed. It traverses a capillary tube having a 10^{-4} inch diameter just as easily as water flows down a large pipe. Another remarkable property of liquid helium is its ability to conduct heat. One cannot, of course, put very much heat into liquid helium without raising its temperature above the transition point, but small amounts of heat energy are transferred through the fluid with the speed of sound.

Thus there seem to be two kinds of liquid helium, one existing above 2.19°K and the other below; these are called *liquid helium I* and *liquid helium II,* respectively. The pathological properties of liquid helium II are now partly understood and provide an interesting illustration of a cooperative phenomenon. One of the many experiments for isolating and exhibiting some of these properties results in what is known as the fountain effect. A vessel ending in a very narrow capillary is immersed in a bath of liquid helium. The vessel has a tube leading from the top. If a small amount of heat is introduced into the central vessel, say by an electric current in a wire or even by light directed onto it, a jet is formed that continues to rise into the air as long as the heat is supplied.

An explanation of this and a number of other peculiar properties is given by the *two-fluid model* of liquid helium. We suppose that liquid helium II is actually a mixture of two different fluids. One of them, called the superfluid, has all the strange properties—large heat conductivity and little or no viscosity—whereas the other is, and is called, the normal fluid. As helium II is warmed toward the transition temperature, the amount of superfluid decreases and the amount of normal fluid increases until at the critical temperature there is no superfluid in the mixture. When any superfluid is present, the mixture conducts heat very well, and the superfluid part flows easily through openings that hold back the normal part. The two-fluid model accounts very simply for the fountain effect. The capillary dipping into the bath of liquid helium can easily be penetrated by superfluid, which tends to fill the vessel above it. When heat is introduced, it changes some of the superfluid into normal fluid, which cannot flow back through the capillary. But superfluid keeps flowing in, the pressure builds up, and the fountain begins.

Now we have to explain why there are two fluids, and this question has agitated quite a lot of first-rate scientific speculation in the last twenty years. It seems that the best way to comprehend the behavior of liquid helium is to think about phonons in a liquid. A liquid at absolute zero is empty of phonons. If we introduce a little heat, phonons appear—the more heat, the more phonons. These phonons collide with each other and act in many ways like gas atoms. In these terms, the superfluid can be considered as the underlying liquid helium itself. The normal fluid does not really consist of atoms at all but of phonons. The phonon gas flows around in the superfluid, so that what was at first thought to be a mixture of two fluids turns out to be a gas superposed on a background fluid, and heat flows through the liquid at the speed of sound because it *is* sound.

It is exceptionally difficult to make a consistent mathematical theory of this cooperative phenomenon. Many things stand in the way, notably the fact that the nature of the law of force between two helium atoms plays a complicating role. Theorists have therefore concentrated a good deal of attention on a simple model that is expected to show some of the same properties as real liquid helium though perhaps not all of them. The model is of a gas of hard spheres that do not exert any of the mutually attractive forces which one knows helium atoms exert but that, nevertheless, repel each other when they bump together. Such a gas would contain phonons, because if one were to make a disturbance in it at one point, some sort of compressional wave analogous to sound would be generated and would travel through it. The quanta of such a wave are phonons. Since the characteristic properties of any liquid are largely due to the way its molecules stick together, the hard-sphere gas is not really a model of liquid helium. The beauty of the model is that it is just near enough reality to demonstrate a few features of the real situation; at the same time, it is just barely possible to discuss the model in mathematical terms, for almost the extreme limit of modern analytical technique is reached in the theory of the hard-sphere gas at low temperatures.

The results of the theory depend very much on quantum mechanics—that is to say, they cannot be substantiated by arguments based on classical physics. Several properties of the two-fluid model of helium II have now been explained, particularly by the brilliant collaboration of the Chinese-American physicists T. D. Lee and C. N. Yang at Columbia University and the Institute for Advanced Study. They found that the behavior of the phonons in the hard-sphere gas corresponds in many respects to the thermal and dynamic properties of the normal fluid in liquid helium; there is every reason to hope that more work in the same direction will lead to an inclusion of the interatomic forces in a realistic way and therefore to a more accurate representation of the experimental facts.

One point about liquid helium is very important. There are in nature two isotopes of helium, one of which, called helium-4, has a nucleus consisting of two neutrons and two protons; this is the common species that we have been discussing. The other is helium-3, with two protons but only one neutron; it is present in nature in very small amounts and must be collected by a long and difficult process of refinement. Chemically, and to some extent physically, the two types of gas are very nearly indistinguishable. Helium-3, being a little lighter, condenses from a gas into a liquid at a slightly lower temperature than helium-4, but the normal fluid properties are very much the same. There are, however, profound differences between these two gases based on quantum-theory phenomena having to do with the statistical distributions of gas molecules among different energies. Because helium-3 obeys the exclusion principle whereas helium-4 does not, the formulas for these distributions are markedly different at low temperatures. There was for many years disagreement as to whether the properties of liquid helium depend very sensitively on the nature of these statistical distribution formulas—the idea of phonons in an otherwise undisturbed fluid applies equally to either isotope.

The crucial test of whether the quantum-statistical distributions play a major part in the explanation of the low-temperature behavior of liquid helium is to see whether liquid helium-3 acts like liquid helium. A few years ago large enough amounts of helium-3 became available for the matter to be investigated. It turned out that helium-3 is normal all the way down to the lowest temperature that has thus far been studied, a few hundredths of a degree above absolute zero. It is therefore apparent that the quantum nature of the gas is of great significance, and this conclusion is built into the theories of Yang and Lee analyzing the behavior of the hard-sphere gas at low temperatures. Whether anything happens to helium-3 as its temperature is further decreased is still not entirely clear. There are some theoretical reasons for supposing that it too will turn into some sort of superfluid if it is cooled down far enough. However, the very low temperatures that have so far been reached reveal no sign of anomalous behavior, and it is quite possible that helium-3 may be an ordinary liquid down to the lowest temperatures that will ever be attained.

PROSPECTS

It is evident that we cannot claim to understand the phenomena of nature until we have shown that we can explain the cooperative behavior of systems containing large numbers of elementary particles in terms of their elementary interactions. In a few places explanations of such behavior are imminent; but to return to the examples with which this chapter opened, any rigorous theoretical interpretation of even such familiar phenomena as boiling and freezing is unknown. Enlightenment moves in two directions in this study. A knowledge of the elementary interactions is necessary for an understanding of the cooperative interactions, but the cooperative interactions themselves furnish insights into the elementary interactions that perhaps can be obtained in no other way.

The whole future of scientific knowledge, considered very broadly, seems to tend to an emphasis of cooperative phenomena. Surely one of the major goals of psychology, and even of the science of politics, is to be able to argue from the known psychological qualities of individuals to an understanding of a crowd, a city, a nation, or a planet. The connection between the two forms of behavior is yet far from obvious. It may ultimately develop that a political organization based on a deep understanding of group behavior is destructive to our dignity as individuals. At the same time, it may be that the increased interactions among individuals resulting from the rise in population, as well as the increased number of ways in which one human life can affect another, will require detailed psychological analysis of a rigorous and predictive character (and even possibly psychological manipulation of us all) in order to keep the peace. This is a touchy and unpleasant subject, however, and no more than an analogy with what we have been discussing.

CHAPTER 9
Lasers

Suppose that an atom has an excited state at an energy just 2 electron volts above the ground state. If the atom is in its ground state and a photon with an energy of 2 electron volts hits it, it will likely absorb the photon and pass into the excited state. A moment later it will emit another 2-electron-volt photon and return to the ground state. The two processes are shown in Figure 9–1a, b. The second is called *spontaneous emission*. Spontaneous emission, which has been mentioned several times before, is the ordinary process of radiation. But what happens if a 2-electron-volt photon hits the atom at an instant when it is already in the excited state? The photon may do nothing at all, or it may drive the atom down to its ground state with the emission of another photon; in the latter case, the two photons will go off together in the same direction and exactly in phase with each other, as shown in Figure 9–1c. This process is known as *stimulated emission*. To illustrate, take a box with reflecting walls, and put into it a large number of excited atoms, all alike. Then introduce a photon of the right energy. It will hit an atom and produce a pair of photons, as in the figure. Then each of these will hit another atom, and so on, and a chain reaction will build up. Further, all these photons will be in phase with their com-

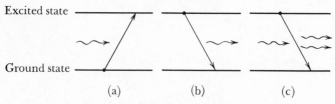

FIGURE 9-1   Absorption (a), spontaneous emission (b), and stimulated emission (c) of radiation.

mon ancestor and therefore with each other. The situation will be quite different from that in an ordinary neon or mercury or sodium vapor lamp, where electron bombardment excites the atoms one by one and they then return one by one to the ground state by spontaneous emission of photons with random directions and phases.

The general idea of an amplifier of electromagnetic waves using stimulated emission occurred, in 1953 and 1954, independently to Joseph Weber of the University of Maryland, Charles Townes of Columbia University, and N. G. Bassov and A. M. Prokhorov in Russia. Because it seemed easiest to handle technically, all of these men proposed to deal with microwave radiation, which is electromagnetic radiation with a wavelength of a few inches. This is about 10^4 times as long as the wavelength of light and can be produced by molecular motions in the same way that atoms produce light. Only Townes went ahead with design studies and construction, and in 1955 he made an amplifier that worked. He called it a *maser*, the word being an acronym formed from the first letters in the phrase "microwave amplification by stimulated emission of radiation."* Since the radiations from all the molecules are synchronized with each other, the output of the whole device is a wave of remarkable purity. A maser need not be used solely as an amplifier. The exciting photon can originate inside it just as well as outside, and as long as the supply of excited atoms can be maintained (this is the

*The rumor that maser stands for "money acquisition scheme for expensive research" is malicious and without foundation.

real trick), the maser will continue to radiate. It is now possible to keep one going indefinitely, and the vibrations from two such maser oscillators can be made to keep time with each other with an accuracy of one part in 10^{11}. A maser oscillator used as a clock would keep time to within about a second in a thousand years.

As soon as a maser had been made to work, people began to think about building a *laser,* in which the microwaves are replaced by light. The difficulties were great, but they were of degree rather than of principle. The worst was that most atoms excited by optical radiation tend to decay by spontaneous emission within 10^{-8} second, and therefore atoms must be energized quickly and simultaneously in a laser for any induced emission to be generated before all the excited atoms are gone. Happily, there are a few substances that can be used in which spontaneous decay is much slower. The obvious way to energize the atoms is with a flash of brilliant light. But such a flash contains light of all colors, that is, all wavelengths. Accordingly, its photons are of all energies, and very few will have just the right energy to cause the desired transition. It is therefore hard at first to see how energy can be transferred quickly enough to the atoms to excite a large fraction of them at the same time.

Between 1958 and 1960, Townes and Arthur Shawlow of the Bell Laboratories examined various schemes for overcoming these difficulties and also for containing the light if any should ever be produced. They also called attention to the interesting optical properties of the ruby. This is a crystal composed of colorless aluminum oxide with a small impurity of chromium that makes it red. Figure 9–2 shows the relevant energy levels of the chromium atom and some transitions that can occur. A broad spectrum of the initial flash is absorbed in transitions to the energy bands labeled F_1 and F_2. The atom then rapidly loses radiation, not by radiation but directly to the surrounding aluminum oxide crystal, and ends up in the state E_1. From E_1 the atom returns to the ground state by emitting a photon of red light. These processes happen in a ruby all the time and give it its characteristic brilliance. It is

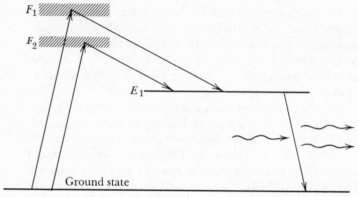

FIGURE 9–2   Some energy levels of a chromium atom in a ruby crystal. Light from a flash tube energizes the electrons into two broad bands of energy levels, F_1 and F_2, from which they quickly drop into the sharply defined and relatively long-lived state E_1. If a photon of the appropriate energy arrives at this point, it induces a transition to the ground state, with the emission of a second photon.

actually creating red light out of the blue and green parts (F_1 and F_2 energy bands, respectively) of the white light that falls upon it.

In July, 1960, Theodore Maiman at the Hughes Research Laboratories made the first ruby laser. The ruby was a synthetic one, pink rather than red, about a centimeter on a side, with two opposite surfaces ground very accurately flat and parallel and then silvered, so that light could reflect back and forth many times without wandering away. On one side the silvering was thinner, reflecting only about 75 per cent of the light and allowing the rest to pass out (Figure 9–3). The reason Maiman got a ruby to "lase" before anyone else was that he tried harder. The gigantic flash of white light with which he energized the chromium atoms delivered 2000 joules of energy to the crystal, enough to lift a heavy trunk onto the roof of a house. About a thousandth of a second after the white flash, red light emerged from the end of the laser. It took the form of several hundred random pulses, each lasting about a millionth of a second, and it was all gone in about another thousandth of a second. At the peak of one of the pulses, the power was about 30 kilowatts coming out of an area of 1 square centimeter. It was probably the brightest light ever seen in the solar system; even the light at the sun's surface radiates only 7 kilowatts per square centimeter.

Further, the light from a laser is very coherent. That is, it consists of photons of very nearly the same wavelength, very nearly in phase with each other, and moving along very nearly parallel lines. Waves that emerge straight out from the silvered surfaces have been amplified by repeated reflection inside; all others have wandered out of the crystal. The beam is so accurately parallel that in the best lasers it spreads out only a few centimeters in a mile. This property makes it possible to bring it to a fine focus at which it is intense enough to punch a tiny hole through a ⅛-inch steel plate or to vaporize any known substance.

Recent pulsed lasers are more powerful than the original

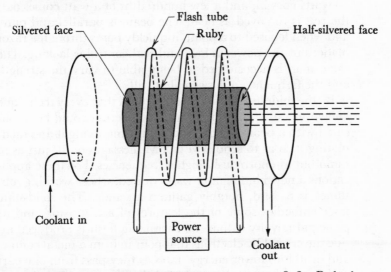

FIGURE 9–3   Ruby laser.

model. In 1962, R. W. Hellwarth and F. J. McClung discovered that if one delays the onset of "lasing" until the excitation of the chromium atoms is completed, the discharge comes out in one giant pulse that lasts a few ten-millionths of a second and reaches a peak power of over 1,000,000 watts.

The pulsed lasers are very impressive, and the pleasure of making and improving them has engaged over a hundred good physicists and engineers for the last year or so, but they have few known uses. The energizing flash pumps so much heat into the crystal and its surroundings that the whole thing has to cool for several minutes before it can be fired again. And it is hard to control and shape such a short-lived explosion of energy. A laser was needed that would produce a steady and continuous beam of pure, coherent, parallel light like the microwave signal from a maser oscillator. Such a device was made by Ali Javan and several collaborators at the Bell Laboratories in the fall of 1960. The laser action takes place in low-pressure neon gas in a tube about a meter long. Energy is supplied continuously through a rather tricky process mediated by helium gas, which is also present. About 50 watts goes in, and a few hundredths of a watt comes out; the rest is removed as heat. The beam is parallel and pure, and when focused to a point, it quickly bores through a razor blade; but the power is low, and real control is lacking. The radio man is accustomed to being able to vary the strength and the frequency of his signal at will.

By the middle of 1963 it could be said that every technique known to radio and microwave engineering could be duplicated with a beam of light, including amplifying it and modulating it so as to send AM or FM messages on it, just as in radio but at enormously higher frequencies. For these applications a new kind of laser was required, whose working substance is a solid, notably gallium arsenide. The solid-state laser embodies some of the features of a transistor, and its principal novelty is that the energizing flash is produced inside the crystal by electrons that pour in from a metal contact and drop into lower energy states as they pass from one part of the crystal to another with slightly different properties.

The power of a laser of this type is only a watt or so at present, but it will presumably go up, and let us consider for a moment what such a controlled beam will do. The outstanding problem of radio engineers today is the overcrowding of the various frequency bands. From the lowest usable frequencies to the highest, transmitters are squeezed in at frequencies so close together that there is much interference, and the situation gets worse every year. If communications can be extended to optical frequencies, this problem will be solved. For one thing, the available range of frequencies will be greatly increased. Furthermore, the message-carrying capacity of a single modulated light beam is enormous. Someone has estimated that a single laser beam could transmit, simultaneously, all the radio, television, telephone, telegraph, and teletype signals that now clutter the wires and air. We could even increase the traffic. Think of every man, woman, and child having his own television program.

The obvious drawbacks to handling all telecommunications with light beams are that they are stopped by clouds and that they do not follow the curvature of the earth. Something will have to be done with pipes underground or with relay satellites like Telstar.

This is in the future. Already laser beams are being used for tracking satellites, for drilling tiny holes in hard substances, and for welding very small objects that resist ordinary welding procedures. The first worthwhile purpose a laser ever served was in eye surgery; a single flash, not too strong, makes a little scar that will attach a loosened retina back onto the choroid surface underneath it. The same technique is under study for instantaneously destroying small retinal tumors. This surgery is accurate and painless, since the flash is over before the eye begins to react.

This book has avoided the applications of physics earlier because they tend to be independent of questions of more fundamental interest. A chapter on lasers has been included, however, because the laser itself is fascinating and it has revolutionized our methods of creating and controlling beams of light. As scientific imaginations catch up with the new possibilities, important experiments will ensue. At the moment of this writing, Professors Townes and Javan are at work in a rural Massachusetts wine cellar, selected because it is in a quiet town and goes down to bedrock, where vibration is minimized, repeating with great accuracy some of the basic experiments that support the special theory of relativity. The preliminary results, already a little more accurate than any previous ones, agree with the theory, and it is estimated that the accuracy can be increased by a further factor of 100. But they are seeking new answers to old questions, and it seems unlikely that anything noteworthy will turn up. The laser will come of age when it supplies the answer to a new question.

CHAPTER *10*

Symmetry in Nature

Let us make a highly unlikely supposition and see where it leads us. Suppose that someday we establish radio or television contact with a being on a distant planet in a remote solar system and that somehow we teach each other a common language so that we can communicate back and forth with perfect ease. A few word meanings would be troublesome to convey, and there is just one pair that we might think about: these are the words *right* and *left*. When one speaks of something being on his right, he is saying that it is on the same side as a particular hand when he is facing forward. If he were turned the other way, it would be on the opposite side of his body, and so the direction in which he is facing must be taken into account when he refers to right and left. Television would be of no help here, because there would be nothing to prevent the receiver from being wired up backward with respect to the transmitter; this situation would result in all the images on the screen being mirror images of the scene picked up by the transmitter. (Interchanging a single pair of wires will produce this effect on any television set.) Alternatively, the terms to be described could be the directions in which one turns a screwdriver to drive in an ordinary screw; the top of the screw moves to the right, and the bot-

159

FIGURE 10–1   Balance, symmetrical right-and-left
but not front-and-back.

tom to the left, as the point moves away from him. Here a
sense of rotation is coupled to a sense of forward motion. If
one stands in front of a screw that is being driven in, it is seen
to turn in the opposite direction. If either the distinction between screw and anti-screw sense or the distinction between
right and left can be made clear, then the other follows easily.

In order to see what the problems of communication are,
let us make a further supposition: that when the perfectly
symmetrical balance shown in Figure 10–1 has both pans
empty, the right-hand side always drops. Then it would be
easy to explain right and left to the being on the distant
planet. We would only have to describe to him in detail how
to make the balance. Since it is symmetrical in its form, we
would not have to distinguish between right and left in re-
laying the instructions. He would construct it, stand back,
and watch the right end drop, and the problem would be
solved. But this, alas, is not what happens. We know perfectly

well that symmetry of structure implies symmetry of function, and we know without building such a device that neither arm would dip. This is exactly why the image of an equal-armed balance is used in public statues as a symbol of judicial impartiality.

It would seem that the universe is truly impartial between right and left. If so, our communications with our remote friend would always contain an element of ambiguity. This much was assumed to be true by the founders of classical physics and their successors who set up the modern quantum theory of fields. It was assumed because there was no evidence to the contrary, because symmetry is regarded as a beauty of nature, and because it is a cardinal principle of people who consecrate their lives to the study of science that nature is beautiful.

In 1956, Yang and Lee were puzzling over a problem that had arisen in interpreting some experiments on elementary particles. They saw that the puzzle could be resolved if the symmetry hitherto observed in nature between right and left were broken. They at once established that no critical experiments on this point had been done in the domain of elementary particles and proposed one or two that could be carried out with some effort and skill. In December, 1956, word came from Washington that a group at the National Bureau of Standards, headed by Mrs. C. S. Wu of Columbia University, had performed an experiment in which nature showed an absolute preference for a left-handed screw over a right-handed one.

The Washington experiment can be described simply as one that is not the same as its mirror image. It is illustrated in a highly schematic manner in Figure 10–2, where the left-hand sketches show the experiment as it was conducted and the right-hand ones show its image as it would appear in a mirror. The first diagram shows a left-handed screw and its mirror image, an ordinary right-handed screw. If some experimental arrangement can be found that does not behave the same as its mirror image, then the difference can be used to distinguish between the two screws, giving a universal specification of right and left.

Suppose that we take a lump of radioactive matter and place it at the center of a coil of wire carrying a current, as in the second diagram. The mirror image is exactly the same, but many atomic nuclei act like little magnets and are capable of being lined up by the horizontal magnetic field generated by the current. Ordinarily, such an alignment does not take place, because the random thermal jostling of neighboring atoms is enough to destroy it; but thermal motions decrease with decreasing temperature, and one can cool the whole sample in its magnetic field to such a low temperature, a few degrees above absolute zero, that the nuclear magnets begin to align. They are represented as pins in the next diagram, which indicates that one end of a magnet is not necessarily indistinguishable from the other. Mirroring reverses the pins, and here is where the radioactivity comes in, for it is used to distinguish between the ends. The first experiments involved nuclei of cobalt-60, which emits beta rays (high-speed electrons) and which is, in addition, more than usually magnetic. The guess was that the beta rays might be emitted either mostly toward the head of the pin or mostly toward its point. The experiment by Wu, Ambler, Hayward, Hoppes, and Hudson, reported in masterly fashion in the *New York Times* of January 16, 1957, showed that they are ejected toward the right. In the mirror image, of course, they would emerge toward the left. Nature therefore distinguishes between a right-handed screw and a left-handed one, for if a left-handed screw is turned in the same direction that the current flows around the loop, its point advances in the same direction that the beta rays are emitted.

Two weeks after the Washington experiment, a second one was announced. F. L. Friedman and V. L. Telegdi of the University of Chicago made a careful statistical survey of a large number of muon decays and found that in a small but significant majority the positron was emitted back in the direction from which the muon came. The muons had come to rest before they decayed, but they had somehow managed to

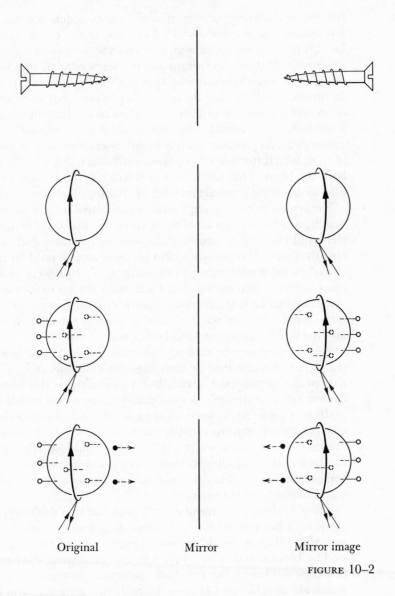

Original   Mirror   Mirror image

FIGURE 10–2

SYMMETRY IN NATURE

remember their original direction of motion. A detailed study has shown that a muon, which has to be positively charged for this particular experiment, is born in the decay of a positive pion and starts out spinning in the sense of a left-handed screw. As it loses headway, the spin begins to drift away from its original line, but the decay into a positron and two (unobservable) neutrinos usually occurs before the spin direction is entirely randomized. Like the nucleus of cobalt-60, the muon emits its positron in a preferred direction along its axis of spin, which turns out to be backward along its path. Thus both the birth of the muon and its death take place in a way that violates the general right-left symmetry of nature.

Returning to our distant friend, what if he is an antibeing, that is, a being composed of antimatter? In his body the nuclei would consist of antineutrons and antiprotons, and the electrons would be positrons, and his laboratory would be an antilaboratory containing anticobalt-60. If this were true (and we have seen earlier that there is no reason to suppose that it cannot be true), then the electric current flowing in his loop would from our point of view be flowing the other way, and the whole argument would collapse.

Is there any way to save it? We could hopefully tell him, perhaps, to be sure that he used negative electrons and positive nuclei in his experiment, but we would see this hope dashed because the only way of distinguishing negative and positive is precisely in terms that must be used to define the charges rather than to identify them. So, if this being is an antibeing, then what we mean by right would be what he means by left, and all communication between us would be completely consistent with itself and in agreement with all experiments he could perform.

There is also a third possibility. It might be that this being's clock runs backward relative to ours. If this were true, then something that we would say is turning to the right would be said by him to be turning to the left, and once more the confusion would begin. But it is quite clear that we cannot communicate sensibly and become friends with someone who is growing younger while we are growing older and who has experienced the end of our communication before it has begun, and so we hope that this possibility is only academic. What is not academic is that nature does not exhibit perfect symmetry between right and left, called a *parity symmetry* in the jargon of the trade, or perfect symmetry between positive and negative charges but does seem to exhibit a combined symmetry, in the sense that if one replaces all particles by antiparticles and at the same time interchanges the signposts of right and left, the world that results is identical with the world with which one started.

These remarks are possibly even more remote from any conceivable area of everyday concern than the other statements in this book, and they have therefore been placed at the end. And yet all of physics is interconnected. When the equations thought to govern interactions of quantized fields were being written down, the writers supposed without further analysis or discussion that the mathematical expressions had to be separately symmetrical with respect to parity and with respect to charge. The decay of cobalt-60 and of muons is governed by weak interactions, and for them it appears that this assumption is not true. Strong and electromagnetic interactions show symmetry in parity and in charge separately, and about gravitational interactions we do not know. Nowadays the mathematics is simpler because it must reflect only a combined symmetry in charge and parity; and parts of the theory, at least, are beginning to agree with experiment. The discovery of Yang and Lee, which seemed at the time to relate to a very minor point in the physicist's picture of nature, thus may have begun something of a Golden Age in the theoretical interpretation of elementary-particle physics. This fundamental asymmetry—or symmetry, but of a deeper kind than anyone had previously considered—is now firmly established at the very basis of our understanding of the theory of weakly interacting fields. There it will stay until it is superseded by some new insight from the physics of the future, which is surely coming though the geniuses who will introduce it may not yet be born and we have no idea what it will be.

Appendix
Index

Appendix

Physics is full of numbers like 21,700,000,000,000,000 and 0.00000000034. These are hard to read, for few people want to count so many zeros or know the names of the numbers containing them. Everyone in science, therefore, uses a certain simple notation, based on the idea of an *exponent,* or *power.* For example, 6^2 means 6×6, or 36, and 10^3 means $10 \times 10 \times 10$, or 1000. The superscript numeral is said to be the exponent of the numeral on the line, or the power to which the number on the line is raised. In 10^3 the exponent 3 is equal to the number of zeros in 1000. With this notation, then, 6500 becomes 6.5×1000, or 6.5×10^3, and the first number in this paragraph becomes 2.17×10^{16}, read "two point one-seven times ten-to-the-sixteenth."

The exponential notation simplifies not only the reading and writing of large and small numbers but also the arithmetic involving them. Suppose that we wish to multiply 1000, or 10^3, by 10,000, or 10^4. We have

$$10^3 \times 10^4 = (10 \times 10 \times 10) \times (10 \times 10 \times 10 \times 10) = 10^7$$

In other words, we just add the exponents. The product of 3.2×10^7 multiplied by 2×10^8 is 6.4×10^{15}, since we multiply the numbers in front of the multiplication signs separately.

Let us now carry out a division. Obviously, if we add exponents in order to multiply, we subtract them in order to divide. Hence

$$10^7 \div 10^4 = 10^{7-4} = 10^3$$

What about $100 \div 1000$? This becomes $10^2 \div 10^3 = 10^{2-3} = 10^{-1}$. But everyone knows that $100 \div 1000 = 1/10$, or 0.1. So it is clear that negative exponents represent numbers between 0 and 1. Since $10^3 \times 10^0 = 10^{3+0} = 10^3$, we see that the factor 10^0 causes no change.

With our new information we can make a table as follows:

$$10^3 = 1000$$
$$10^2 = 100$$
$$10^1 = 10$$
$$10^0 = 1$$
$$10^{-1} = 0.1$$
$$10^{-2} = 0.01$$
$$10^{-3} = 0.001$$

and so forth. The second number in the first paragraph thus becomes 3.4×10^{-10}, and the product of the first two numbers is

$$2.17 \times 10^{16} \times 3.4 \times 10^{-10} = 2.17 \times 3.4 \times 10^{16-10}$$
$$= 7.378 \times 10^6$$

To have obtained this result by longhand would have been a wearying operation.

Index